Monitoring ecological change

MONITORING ECOLOGICAL CHANGE

Ian F. Spellerberg

Senior Lecturer in Biology
University of Southampton

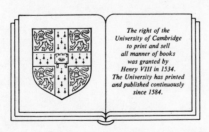

The right of the
University of Cambridge
to print and sell
all manner of books
was granted by
Henry VIII in 1534.
The University has printed
and published continuously
since 1584.

CAMBRIDGE UNIVERSITY PRESS

Cambridge

New York   Port Chester   Melbourne   Sydney

Published by the Press Syndicate of the University of Cambridge
The Pitt Building, Trumpington Street, Cambridge CB2 1RP
40 West 20th Street, New York, NY 10011-4211, USA
10 Stamford Road, Oakleigh, Melbourne 3166, Australia

© Cambridge University Press 1991

First published 1991

Printed in Great Britain at the University Press, Cambridge

British Library cataloguing in publication data

Spellerberg, Ian F.
Monitoring ecological change.
1. Ecosystems
I. Title
574.5

Library of Congress cataloguing in publication data

Spellerberg, Ian F.
Monitoring ecological change / Ian F. Spellerberg.
p.   cm.
Includes bibliographical references (p.   ) and index.
ISBN 0 521 36662 3 (hardback)   ISBN 0 521 42407 0 (paperback)
1. Environmental monitoring.   I. Title.
QH541.15.M64S73   1991
574.5'22 – dc20

ISBN 0 521 36662 3 hardback
ISBN 0 521 42407 0 paperback

Contents

    4   ELEMENTS OF ECOLOGY AND ECOLOGICAL
          METHODS   63
          Introduction   63
          Populations   63
          Communities and ecosystems   69
          Ecological methods   77
          Data collection and recording   88
          References   90

    5   BIOLOGICAL INDICATORS   93
          Introduction   93
          Plant and animal indicators   94
          Detectors and exploiters   98
          Accumulators   106
          Status of biological indicators in monitoring programmes   108
          References   109

    6   DIVERSITY   112
          Introduction   112
          Number of species, species composition and abundance   112
          Species diversity indices   117
          Variation in species diversity   123
          Application of diversity indices   125
          References   129

    7   SIMILARITY   131
          Introduction   131
          Community similarity indices   131
          Application of community similarity indices   136
          Ranking, classification and ordination   136
          References   140

    8   ENVIRONMENTAL AND BIOTIC INDICES   142
          Introduction   142
          Environmental indices   142
          Wildlife indices   144
          Biotic indices   145
          Development of biotic indices   157
          References   158

    9   BIOLOGICAL VARIABLES, PROCESSES AND
          ECOSYSTEMS   160
          Introduction   160

Foreword

In everyday life, nobody would think of buying a car without speedometer, fuel gauge, temperature indicator, and warning lights to draw attention to various problems like a loss of oil pressure. These are all monitoring devices, telling us how our vehicle is working and allowing us to avoid the catastrophe of engine seizure or the annoyance of running out of fuel in the rain on a cold Sunday night.

The environment is vastly bigger and more complicated than any man-made machinery and the consequences of system failure are more disastrous, yet we commonly make major decisions about its management with scarcely any adequate indicators of its state, trends or response to our collective impact. The result, all too often, is unforeseen damage, diminished productivity, and a loss of what we now call 'sustainability' – 'the ability to meet the needs of today without jeopardizing the ability of future generations to meet their own needs'.

Monitoring is the process by which we keep the characteristics of the environment in view. It provides the essential data on how systems are changing and how fast. It provides the essential feed-back loops to management, so that we can adjust what we are doing and get the best out of the system. We need it, whether we are concerned to regulate pollution, manage fisheries, sustain soil fertility or look after nature reserves.

But what should be monitored, out of all the bewildering complexity of nature? It is not possible to measure everything – choice is imperative. Very often that choice falls on physical attributes like temperature or chemical variables like the concentration of a key nutrient or of a significant pollutant. Such things are relatively easy to measure and the resulting tables of data look impressive. However, they tell us rather little about the response of ecosystems or species.

Biological monitoring starts at the other end. Its logic rests on the fact that living organisms integrate the impact of many variables and that their biological efficiency, productivity or balance within the ecosystems they compose indicate the overall health of the system. Lichens growing on tree trunks are highly sensitive indicators of air pollution: the state of fresh waters can be judged from the faunas they support: the pattern of tropical vegetation can give eloquent expression to its history. And since we commonly manage the environment in order to sustain particular biological features, the direct surveillance of its biological characteristics is likely to be the best way to establish whether the assumptions behind our management plans are valid.

This book is intended to introduce the concepts and practices of biological monitoring to the student and the general professional reader. It is deliberately illustrative rather than a comprehensive account of the innumerable biological monitoring systems in the world. Instead, it explains what kind of biological monitoring might be used when, for what purpose – and how the results can be analysed, evaluated and applied.

There are no short cuts in the sphere of environment, and biological monitoring is not a miraculous cure-all. Its results can be as difficult to evaluate as any other scientific data about complex systems. But it is a tool we are using increasingly in today's world, and has an important contribution to make in safeguarding tomorrow's environment. If this book teaches more people to understand and use it, it will make an important contribution to the world's future.

Martin W. Holdgate   IUCN – The World Conservation Union
Director General

Preface

For some readers, the title of this book will be considered to be misleading. There are many books on biological monitoring, some general and some specialized. Most of those books define biological monitoring in the context of environmental quality and are therefore biassed towards monitoring the effects of pollution. Few books on biological monitoring consider monitoring in a broader and ecological sense, that is in connection with studies of changes in biological communities, ecosystems and the status of species. Here, we assess how changes in biological communities, ecosystems and populations have been monitored. We also assess the role of single organisms in monitoring environmental change. Some of the changes observed in biological communities and ecosystems are natural and some are caused by man. The emphasis here is not necessarily on pollution or even the effects of physical disturbances brought about by man's exploitation of nature and natural resources. The emphasis throughout the book is on the assessment of biological and ecological monitoring programmes, particularly with regard to conservation.

The reason for writing this book is because I believe that biological monitoring is important for several reasons, including the following.

1. The world's living resources are being depleted all the time and impacts from developments are affecting the quality of those living resources. If sustainable development is to be a world objective then we need to monitor changes in those living resources as a basis for modelling strategies for sustainable development.
2. Biological and ecological monitoring has an important role in management of plant and animal populations for conservation. Without monitoring changes in natural communities, without monitoring changes in the status of species, without monitoring the effects of

habitat loss, then we have little on which to base good conservation practices.

3. Studies of land use and landscape change will come to rely more and more on good ecological monitoring techniques.

4. The quality of our water, the air and the soil can be monitored through the use of indicator species and indicator communities, far more successfully than by chemical monitoring alone.

5. There have been few long-term ecological studies and consequently we know very little about natural long-term processes of ecosystems. Biological monitoring programmes have an important role in our understanding of those processes and provide essential baseline data in studies of the effects of environmental impacts.

6. Biological monitoring has important applications in assessing methods of controlling pests and diseases of concern to agriculture and forestry.

Environmental monitoring has not been totally without initiatives. Since the early 1970s there have been some major world developments and initiatives in monitoring and surveillance programmes. Rapid advances in information technology have helped those developments but advances in technology do not seem to have been exploited as much as they could have been. Despite its importance, funding and support for biological and ecological monitoring has been minimal. Perhaps this is because monitoring pollution, energy and other resources is considered more important. I believe there is an urgent need for the establishment of many more long-term biological studies. Surely there is no doubt about the desperate need for more well-planned and well-administered monitoring programmes. The few long-term biological monitoring programmes which exist today provide data which become more valuable as the monitoring programme continues. Unfortunately, identifying the need for a biological monitoring programme is relatively easy compared to the initiatives which are required to establish those programmes. All too often the monitoring comes too late or is poorly planned and consequently cannot provide the data which can help in conservation or management of natural communities. Perhaps more education and training in biological monitoring could redress these circumstances and perhaps the establishment of centres of biodiversity could help to initiate such programmes.

Finally, I should comment on the limited examples of statistical analysis included amongst the text, for which I offer no apologies to the computer-knowledgeable reader. One of the aims was to write not only an informative text but a reasonably self-sufficient text. I have therefore included some basic statistical analysis with some worked examples but knowing full well that there are many comprehensive statistical texts.

Even better, computing facilities and the range of statistical packages will have superceded my limited examples. While I encourage everyone to make use of computing and the range of time-saving statistical packages, I know that some readers will not always have computing facilities available at a time when they need them. It is for those occasions that I have included what I hope to be basic, but useful, worked examples.

Ian Spellerberg   Southampton

Acknowledgements

This book has been long in thought and long in preparation. It is inevitable, therefore, that over the years many of my colleagues have influenced my thinking about biological and ecological monitoring. To say that I am grateful for their help and encouragement seems to be far too inadequate. I am particularly pleased to acknowledge the following who have all contributed in some way to improving various parts of the manuscript: M. C. Acreman (NERC, UK), Roger Bamber (Fawley Marine Biological Laboratory, UK), Brian Birch (Southampton University), John Cairns Jr. (Virginia Polytechnic Institute and State University), Raymond Cornick (Southampton University), Eric Cowell (BP International), Mike Fenner (Southampton University), David A. Flemmer (US Environmental Protection Agency), David Given (DSIR, New Zealand), Jeremy Greenwood (BTO), Jeremy Harrison (WCMC), J. M. Hellawell (Nature Conservancy Council, UK), M. Holdgate (IUCN), Peter Hopkins (Department of Agriculture, Food & Fisheries, Scotland), S. M. House (ANU, Australia), M. J. Hutchings (University of Sussex), I. G. C. Kerr (Tussock Grasslands & Mountain Lands Institute, NZ), Mike Ladle (Institute of Freshwater Ecology, UK), Gene E. Liken (The New York Botanical Garden Institute of Ecosystem Studies), Danielle Mitchelle (GEMS), David Norton (University of Canterbury, NZ), Mary E. Palmer (previously with the Nature Conservancy, USA), Robin Pellow (WCMC), Franklyn Perring (UK), Derek Ratcliffe (UK), P. Robertson (The Game Conservancy Trust), Peter Rothery (British Antarctic Survey), A. A. Savage (UK), John Shears (BAS), Roger Smith (University of Newcastle upon Tyne, UK), K. Spellerberg (Southampton), L. Roy Taylor (UK), David Truager (US Fish & Wildlife Service), Paul Vickerman (Southampton University), Richard White (Southampton University), Anthony J. Whitten (UK), Douglas A. Wolfe (US National Oceanic & Atmospheric Administration), John Wright (Institute of Freshwater Ecology, UK) and B. K. Wyatt (Environmental Information Centre, The Institute of Terrestrial Ecology, UK).

Part A

ENVIRONMENTAL AND BIOLOGICAL MONITORING

1

The science and art of monitoring

Introduction

THE ANTARCTIC REGION, including the great ice-covered land mass and the biologically rich oceans, would appear to be a region of the world which is relatively safe from exploitation and sufficiently remote not to be harmed by pollution. The Antarctic region might also be considered to be one of the last locations on earth where there was a need to undertake any kind of monitoring or surveillance of the wildlife. However, in one survey of the impacts of man's activities in the Antarctic, a very wide range of potential impacts on the terrestrial and marine environment were identified (Table 1.1).

For over 200 years the Antarctic oceans have been exploited for whales, fish and plankton. Pollutants from the industrialized world have reached and penetrated the Antarctic ecosystems (Sladen, Menzie & Reichel 1966) and the operational activities of Antarctic research and exploration have had their deleterious impact on the coastal populations of birds and mammals (Bonner 1984). Already one supply ship (the Bahia Paraiso) has spilled over 200 000 gallons of fuel in Antarctica and we know very little about the way in which oil pollution affects ecosystems in polar regions, let alone temperate coastal regions (see p. 281).

Over the last 30 years, an Antarctic Treaty (1959) with its 12 signatory nations has afforded little protection to the biota of this region of the world. Today, however, the attraction of possible extractable minerals and oil poses a new threat to the wildlife and environment of Antarctica. The present Antarctic treaty includes no specific provisions on mining and the Convention on Regulation of Antarctic Minerals Resource Activities (CRAMRA), which would have allowed exploration for minerals and controlled mining, has not been successful because of a lack of signatories.

Table 1.1. *Some environmental impacts (deliberate, incidental or accidental) in the Antarctic. (Some very unlikely impacts and impacts of negligible severity have been ignored.)*

Terrestrial
(including inland waters)
Habitat destruction/modification
Destruction/removal/modification of biota, fossils, ventifacts, etc.
Modification of vital rates of biota (disturbance to production and/or growth)
Modification of distribution of biota
Introduction of alien biota
Pollution by:
    biocides and noxious substances
    nutrients (eutrophication)
    radionuclides
    electromagnetic radiation
    noise
Modification of thermal balance of environment
Aesthetic intrusion

Marine
(a) Shoreline/enclosed waters/benthos
Habitat destruction/modification
Destruction/removal/modification of biota
Modification of vital rates of biota
Pollution by:
    biocides and noxious substances
    nutrients
    radionuclides
    inert materials (dumping)
    noise
    heat

Atmospheric
Pollution by:
    sulphur oxides
    nitrogen oxides
    carbon monoxide
    carbon dioxide
    hydrocarbons
    radionuclides
    dusts
    microbiota
    electromagnetic radiation
Ozone: local excess at ground level, depletion in stratosphere

From Benninghoff & Bonner (1985).

An Antarctic conservation strategy is now being prepared; the draft strategy has been based on the results from several working groups of the International Union for the Conservation of Nature and Natural Resources (IUCN) and work of the Scientific Committee on Antarctic Research (SCAR), one of the major environmental units of the International Council of Scientific Unions (ICSU). Wide consultation is currently taking place and there is now the possibility, or at least indication, that long-term biological monitoring will become an integral part of research in the Antarctic. There are also indications that biological monitoring may also become routine in relation to assessments of large-scale impacts such as the construction of crushed rock airstrips. Interestingly enough, Environmental Impact Assessments or EIAs (see Chapter 15) have been carried out in Antarctica since 1973/74 (Parker & Howard 1977) and more recently there have been formal and detailed evaluations of the procedures for EIAs in Antarctica. One such assessment by Benninghoff & Bonner (1985) has drawn attention to the importance of monitoring key indicators of environmental change (see Chapter 15 and discussion of indicators in Chapter 5).

In most cases, only the minimum of environmental assessment has ever been undertaken by any country prior to developments and construction in the Antarctic. The British Antarctic Survey has at least completed a general environmental evaluation on a proposed airstrip and has responded to comments received from wide consultation. The report (NERC 1989) on the proposed construction of an airstrip at Rothera Point on Adelaide island is general in its approach but it does mention the completed biological baseline work and the usefulness of post-development monitoring of ecological processes. The need to establish detailed plans of both the logistics and science of that monitoring, which are detailed in the report, can not be emphasized enough (see Chapter 10).

Biological research has taken place in Antarctica for many decades and during the 1960s there was a considerable increase in that research alongside a much increased support for environmental, especially geophysical, investigations. Antarctic biological research has included a wide programme of activities, some of which have been directed at population ecology of mammals and birds. For example, surveillance and census of Seals, Penguins and other birds has been undertaken for many years but few results from that surveillance and census work will make any significant contribution to any Antarctic conservation strategy. There has been a missed opportunity of which I had some small personal experience.

During the 1960s I was a member of a research team in Antarctica and part of my research included an analysis of the population ecology of the Adelie Penguin (*Pygoscelis adeliae*) and the McCormick Skua (*Catharacta maccormicki*). The populations of these two birds were occasionally recorded in the area of Cape Royds, Ross Island, as part of an on-going surveillance programme. My field work over three years made a small contribution to the population records which had been kept before the 1960s and which were kept for many years afterwards.

At that time, there seemed to be little concern about the potential value of data from the census of those birds, especially where it is to be undertaken on a systematic basis. Although we saw the value of those records as providing a 'watchful eye' on the status of the populations, no one seemed to be sure of any long-term objectives of the surveillance programme. The recording was not administered so as to ensure continuity, records were not kept in a central depository and recording methods were not uniform. Many years later, the value and importance of a long-term ecological monitoring programme on those two species in the McMurdo Sound area is obvious. Physical disturbances, pollution and other perturbations seem to have affected the populations of those and other birds in McMurdo Sound as well as populations of other bird species around the coastal regions of Antarctica. Without long-term data from good monitoring programmes, the ecology and population dynamics of many species remain unstudied and therefore the extent and nature of the effects of pollution and physical disturbance can not easily be assessed. It is easy to comment with hindsight but I mention these experiences because they 'set the scene' and introduce some fundamental aspects of monitoring which are discussed throughout the book.

Examples and applications

Long-term studies and monitoring

Although not all long-term biological and ecological studies could be said to be examples of monitoring, there are nevertheless characteristics which are common to both monitoring and long-term studies. The applications of long-term monitoring studies has often been discussed over the last two decades, notably in 1970 at the Marine Pollution Conference in Rome and then again in 1977 in the USA where a recommendation was made to the National Science Foundation that long-term monitoring of ecosystems should be funded and supported (see Chapter 2). Obviously, the decision about the most appropriate time scale for a monitoring programme needs to be linked to the objective of the monitoring activity. The importance of linking objectives to time scales can usefully be shown in a conceptual manner (Fig. 1.1), a manner which also emphasizes an important aspect of monitoring, that is it is difficult to identify natural changes in the absence

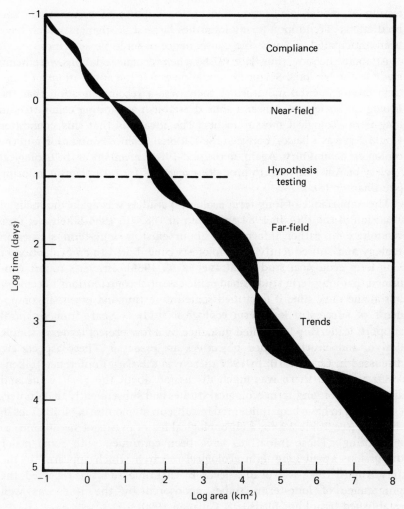

Fig. 1.1. Conceptual relationships of scale in space and time for different categories of monitoring. The temporal scale represents the duration of sampling (with respect to each perturbation) and the spatial scale represents the potential sampling area. Redrawn from Wolfe & O'Connor (1986).

of long-term data. The conceptual approach shown in Fig. 1.1 is based on that aspect of monitoring which deals largely with environmental quality (Wolfe & O'Connor 1986, Wolfe *et al.* 1987) and three main aspects of that monitoring have been identified, viz. compliance monitoring, hypothesis testing and trend monitoring. Compliance monitoring attempts to ensure that activities are carried out to meet statutory requirements. Hypothesis testing or model verification checks the validity of assumptions and predictions. Trend monitoring identifies large-scale changes which have been anticipated as a possible consequence of multiple activities.

At about the same time (late 1970s) when recommendations were being made to the National Science Foundation about the applications of long-term environmental monitoring, there was a recommendation that the Ecological Society of America should establish a standing committee on long-term ecological measurements. The idea was that this committee would serve as a liaison between the National Science Foundation and the ecological community. Again, in the mid-1980s, members of the Ecological Society of America were to promote a network for rare plant monitoring (see Chapter 16).

The importance of long-term ecological studies was again the focus of attention in the USA in 1984 and again in 1987. In 1984 Likens set up a committee to gather information about existing long-term studies in ecology and subsequently 100 ecologists contributed to an evaluation of long-term ecological studies (Strayer *et al.* 1986). Strayer's report confirmed that long-term studies had made essential contributions to ecology but at the same time it identified some interesting and essential components of successful long-term ecological studies. Apart from financial support, leadership (dedicated guidance by a few project leaders), simple design, and clearly defined objectives are essential. These aspects are discussed in Chapter 10. In 1987 there was the Cary Conference (Likens 1989) at which there was much discussion about the great value and applications of long-term ecological studies and subsequently that Conference seems to have been influential in subsequent monitoring initiatives in the USA (see Table 2.1 for a chronological list of events of significance to monitoring). Those initiatives have been continued with some major discussions about long-term ecological research which appeared in the July/August 1990 issue of *BioScience*. By that time, so it was reported, the programme of long-term studies sponsored by the NSF was well established (Franklin, Bledsoe & Callahan 1990).

The British Ecological Society, the oldest ecological society in the world (established in 1913), has also, from time to time, supported the case for long-term studies of ecosystems. At a recent BES workshop held for young ecologists, long-term ecological monitoring was considered to be one of three main research priorities (Hassell 1989). But lack of financial support has prevented the realization of a few significant ecological monitoring initiatives in Britain. A former director of the NERC Institute of Terrestrial Ecology (ITE) recently drew attention to the fact that the importance of long-term research and monitoring, although recognized was not being given sufficient support and commitment from the research councils

(Jeffers 1989). There have been, however, some encouraging initiatives, such as in 1984 when there was the establishment of an Ecological Data Unit within ITE. It is proposed that that Unit will promote the development of a monitoring programme.

Elsewhere in the world the applications of long-term ecological studies and monitoring have begun to be recognized. In New Zealand, for example, a Symposium was held in 1988 on environmental monitoring with emphasis on protected natural areas. Time and time again, speakers at that symposium argued for more support of environmental monitoring programmes while at the same time noting that very little biological monitoring, especially marine biological monitoring, had previously taken place (Craig 1989).

VARIABLES AND PROCESSES USED IN BIOLOGICAL AND ECOLOGICAL MONITORING

What are the biological and ecological variables and processes used in monitoring? The possibilities are very large indeed and it would seem best therefore to give a selection of examples of the more common kinds of biological and ecological variables which could be used in monitoring (Table 1.2). It can be seen from Table 1.2 that biological monitoring programmes have made use of a wide range of variables at different levels; at the ecosystem level through to populations and habitats to the physiological and cellular attributes of specific organisms.

When thinking about a biological monitoring programme we would normally be referring to a programme which had been planned ahead using appropriate variables. Interestingly enough, some biological monitoring can be done retrospectively using historical data. One example of historical monitoring is pollen analysis (Davis 1989), whereby an analysis of the relative proportion and incidence of pollen from different plant species found in deposits of peat or other substrata, together with other souces of information, can be used to construct a description of ecological processes which have occurred over the last few thousand years (Fig. 1.2).

Pollen analysis was one of several techniques used by Engstrom, Swain & Kingston (1985) in a fascinating study of the limnological history of Harvey's Lake, State of Vermont in North America. Today, phytoplankton of that lake is dominated by the blue–green alga *Oscillatoria rubescens* and it is that alga which has left behind pigments in the sediments. These sedimentary pigments were used as indicators of changes in primary production and a detailed, accurate chronology was made possible with the combined use of ^{210}Pb, ^{137}Cs, ^{14}C dating as well as stratigraphy of pollen

Table 1.2. *Biological variables and processes used in monitoring and surveillance*

A. Variables

Biomass
Area of cover or percentage cover[a]
Production
Amount of dead material, litter[a]
Vegetation structure[a]
Lichenometric studies[b]

Species lists (species composition)[a,b]
Species richness
Species diversity
Species frequency[a]
Proportion of all samples in which a species occurs
Occurrence of 'indicator species'
Occurrence of rare species

Phenology of selected species[a]
Spatial patterns of distribution
Population density[a]
Relative abundance of predators and prey
Trophic position

Population age class distribution
Diameter (of trees) at breast height[a]
Birth, recruitment and death rates
Size[a]
Growth rates
Reproductive state[a]
Number in reproductive condition[a]
Plants in flower[a]
Size of breeding colony[b]
Chemical content of living and dead material
Soil structure and composition[b]

B. Processes

Productivity
Litter accumulation
Decomposition
Consumption rates
Carbon, nitrogen fixation
Respiration
Colonization
Succession
Bioaccumulation

This list comes from a wide source: those marked [a] have been proposed by Palmer (1986) for rare plant monitoring (see Chapter 16) and those marked [b] are some of the parameters proposed for monitoring operational impacts of research stations in the Antarctic (Benninghoff & Bonner 1985).

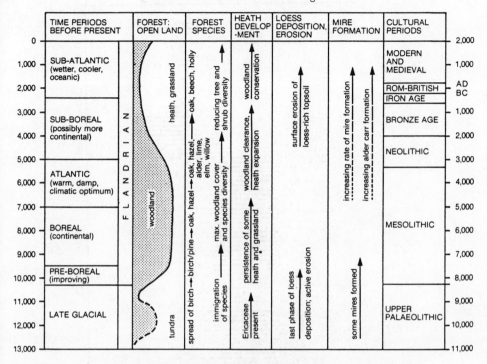

Fig. 1.2. Post-glacial changes in the New Forest environment based on data from pollen analysis, radiocarbon dating and other sources. Reproduced from Tubbs (1986) with kind permission of the author and Collins.

and sawmill waste deposits. It was found that two periods of increased sedimentary anoxia (1820–1920 and 1945 to present) could be attributed to sawmill wastes and later to increased levels of nutrients coming from dairy wastes.

Monitoring the effects of agricultural chemicals on organisms has also been done retrospectively. One classic example, which commenced in 1966, was Ratcliffe's study of thinning of Peregrine Falcon (*Falco peregrinus*) eggshells (Ratcliffe 1980). Rather like a detective, Ratcliffe examined museum collections of eggs taken from various regions in Britain and calculated an eggshell index based on the weight of the egg divided by the length times breadth. The change observed in the shell index (Fig. 1.3) from 1947 onwards has since been shown beyond doubt to have been caused by accumulation of DDT in Peregrine tissues.

Ratcliffe had to use material which was not collected specifically for monitoring variations in eggshell thickness. On hindsight and bearing in

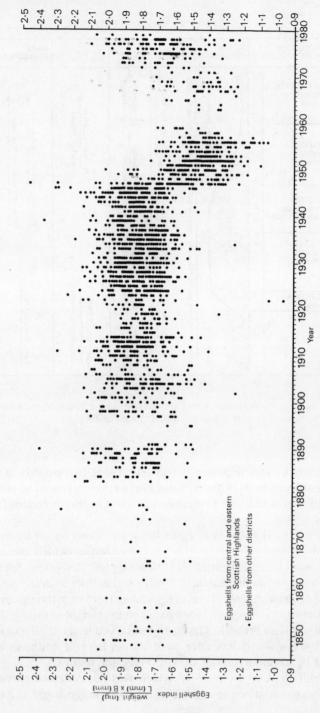

Fig. 1.3. Decrease in eggshell thickness for Peregrine Falcons. From Ratcliffe (1980). Reproduced with kind permission of D. Ratcliffe and T. & A. D. Poyser Ltd.

mind the previous rapid growth in the use of agricultural chemicals (for example, the US total production of DDT from 1944–1968 was 1225 million metric tons (SCEP 1970)), a monitoring programme could have been initiated to look at the effects of those chemicals on non-target organisms. That is, the data could have been sampled in a representative manner and assembled in a systematic manner, thus providing a sound basis for analysis. Much the same could be said for the many agricultural and horticultural chemicals currently in use today.

Other fascinating examples of historical monitoring have come from research carried out by the Monitoring and Assessment Research Centre (MARC) and include one particularly interesting investigation on the accumulation of heavy metals in Deer antlers (MARC 1985). This and other examples are described in detail in the following sections.

The importance and value of biological monitoring

Interest in the changes in populations of plants and animals (especially birds) is both widespread and topical. Not only is there an interest in changes in populations of plants and animals but there is also concern about declines in populations. Pollution, loss of habitat, disturbance and changes in land use may all be implicated in the decline of populations but cause and effect can not easily be demonstrated without data from monitoring and surveillance studies. For without long-term studies it is difficult to distinguish between natural changes and those changes caused directly or indirectly by pollution and other impacts.

Surprisingly, perhaps, we know very little about natural, temporal behaviour of plant and animal populations. We know even less about the long-term processes in ecosystems, especially undisturbed ecosystems. There are virtually no undisturbed ecosystems left on this earth but there have been some attempts to establish long-term monitoring programmes in relatively undisturbed areas. The information from these programmes becomes more and more valuable with time not only in terms of understanding natural processes but also in terms of providing a baseline for comparisons when disturbance or perturbation occurs.

The four major classes of ecological phenomena which were identified by Strayer et al. (1986) as being appropriate for long-term ecological studies, clearly illustrate the value of monitoring by way of long-term ecological studies. The four classes were as follows: 1, slow processes (ecological phenomena such as forest succession and some vertebrate population cycles which occur on scales longer than the traditional three year funded research or indeed longer than a human life time); 2, rare events (perturbations such as fires and outbreaks of pests or disease); 3, subtle processes (ecological processes where the year-to-year variance is larger than the long-term trend such as found in the biogeochemistry of an aggrading forest (Likens 1982)); 4, complex phenomena (intricate ecological relationships in biotic communities).

The reference by Strayer *et al.* (1986) to slow processes is both important and interesting because there are some ecosystems which function very slowly indeed. Polar plant communities are very slow growing and some plants in the Arctic, for example, may take as long as three years for a flower to bloom and set seed. When thinking about monitoring effects of pollutants in ecosystems, such as radionucleides in grassland communities, we may all too easily forget that there is considerable variation in the rate at which different ecosystems function. Recent concern has been expressed at the high levels of pollutants in Arctic ecosystems and we may have to think very carefully how to monitor pollutants in ecosystems which function very slowly indeed. In Chapter 15 we mention time-scales with reference to the difficulties of assessing when a damaged ecosystem has recovered.

Some biological and ecological monitoring programmes have immediate value with regard to management of natural resources. For example, one good, early example of biological monitoring and one which was established with much foresight was Hardy's continuous plankton recorder study (Hardy 1956). In 1929 Hardy proposed to survey and monitor the distribution and movements of plankton in the North Sea with the object of relating the results to movements of fish population studies. The work of surveying plankton, conceived many years ago by Hardy, is now continued in a less than systematic manner despite important applications in relation to population dynamics of commercially valuable fish.

Biological and ecological monitoring has and continues to play an important role in conservation of species, natural communities and landscapes. World organizations have, for example, been established to monitor population levels and the distribution of endangered plant and animal species. Monitoring changes in land use and the effects of loss of habitats and effects of isolation of populations are other important applications of biological monitoring in conservation.

There have been some most interesting developments in the methods used for insect pest monitoring programmes but equally interesting is the behavioural and ecological information which has come from the long-term studies of insect pests. However, as we shall see in Chapter 3, funding even for this kind of monitoring has been reduced. By way of contrast, there seems to be continued support for pollution monitoring.

Chemical monitoring of pollution has become very highly developed

Table 1.3. *Detection limits and analysis time for some pollutants*

Type of pollutant	Analytical technique	Detection limits	Analysis time[a]	Some applications
Volatile organic compounds	GC/MS with MSD	10 µg/l 50–200 µg/kg	50 min	Waters Sediments
Phenols and nitrocompounds	GC/MS with MSD	25 µg/l	1 h	Waters and wastewaters
Amines	LC/MS	100 ng	30 min	Soil and water samples
Pesticides	GC/MS or GC/ECD	1–10 ng/l 2/10 µg/kg	30 min	Liquids Plants, animals
Polychloro phenols	HPLC	25 µg/l	35 min	Wastewaters
Polyaromatic hydriocarbons	HPLC	10–25 µg/l 0.2–0.5 mg/kg	30 min	Waters Soil samples
Halocarbons	GC with headspace sampler	25 µg/l	20 min	Waters and wastewaters
PCBs	GC/ECD or GC/MS	2 µg/kg 10 ng/l	30 min	Waters Soil samples
Heavy metals	DPP	0.1–5 µg/l	45 min	Waters

From Orio (1989) with permission of Academic Press.

[a]Only for the instrumental step of the analysis.

GC: gas chromatography; MS: mass spectrometry; MSD: mass selective detector; LC: liquid chromatography; HPLC: high pressure liquid chromatography; ECD: electron capture detector; DPP: differential pulse polarography; AAS: atomic absorption spectrophotometry; PCBs: polychlorinated biphenyls.

with the use of very powerful computers and sophisticated technology. It is now possible to detect very small quantities of some pollutants in a relative short space of time (Table 1.3). It would appear therefore that biological monitoring has become redundant but nothing could be further from the truth. Chemical monitoring tells us what is there but it does not tell us what the effects are, especially the long-term effects on ecosystems. For this reason, biological monitoring has a very valuable and interesting role to play in monitoring pollution. The use of living organisms to monitor pollution has become very sophisticated and there has been some recent exciting advances in the methodology for automated biological monitoring of pollution as well as monitoring the quality of drinking water (see, for example, Gruber & Diamond 1988).

The use of biological indicators or bioindicators in monitoring programmes has been of particular interest to the International Union of Biological Sciences (IUBS) who at their 21st General Assembly in 1982 initiated The Bioindicators Programme in Ottawa. As well as recommending that the IUBS organize biological monitoring programmes, that Assembly advocated biological indicators in monitoring by way of the following.

1. Encourage scientific and national bodies to develop and improve methods for monitoring environmental change.
2. Promote interdisciplinary and international cooperation in standardizing methods.
3. Encourage the exchange of research results amongst laboratories of the world.
4. Support conferences dealing with bioindicators at cellular, individual, population and ecosystem level.

Pollution knows no political boundaries and therefore we might expect cooperation and collaboration between countries when it comes to monitoring the effects of pollution. European and Scandinavian countries have often discussed collaborative monitoring projects but it was only as recently as 1979 that the Convention on Long-range Transboundary Air Pollution was signed within the framework of the European Commission. This cooperative monitoring programme, which came into force in 1983, should provide regular information on selected air pollutants from about 90 monitoring stations.

Biological monitoring within the European Economic Community (EEC) has not yet been taken very seriously by member governments despite the lack of a Community-wide mechanism for protection of important wildlife areas and despite claims to the effect by the EEC that the Community is becoming more 'environmentally aware'. In August 1988, however, there was a proposal for a Council Directive on the protection of natural and semi-natural habitats of wild fauna and flora. That proposal did mention monitoring in article 24.

1. Member states shall take all necessary measures to ensure the monitoring of the biological communities and the populations of the species specified in accordance with Annex 1 and in the areas classified under article 5. Member states shall send the Commission the information resulting from monitoring, so that it may take appropriate initiatives with a view to the coordination necessary to ensure fulfilment of the objectives of the Directive.

2. The Community and Member states shall cooperate to ensure consistency of monitoring and measurement methods.

The next few years will see whether or not the European Community does accept the need for Community-wide protection of its wildlife and the need for monitoring that wildlife.

If for no other reason, there is an overwhelming case for biological monitoring in Europe because of the European Economic Community's current ambitious £35 billion plan to encourage development in the poorest regions of Europe. The World Wide Fund for Nature (wwf) and the Institute for European Environmental Policy (ieep) have warned that governments are hastily assembling major but possibly unviable schemes, the impacts of which have not been considered sufficiently. Certainly the long-term biological and ecological impacts of most of these major projects have not been considered; often short-term economic advantages and political gain override environmentally damaging schemes. An example of these major projects is afforestation (with eucalypts) of some regions of Portugal, regions not suitable for afforestation. Such a major project will have devastating, long-term damage on the biology of the area. No biological monitoring programme has been planned for these areas, or at least monitoring change in habitats is given very low priority. Clearly, protection of important habitats and also monitoring is required so that impacts of such major projects can be assessed in the future.

The value of biological and ecological and ecological monitoring and surveillance programmes can perhaps be summarized by the following objectives.

1. Monitoring as a basis for managing biological resources for sustainable development and resource assessment.
2. Monitoring so that ecosystems and populations can be managed and conserved effectively.
3. Monitoring land use and landscapes as a basis for better use of the land; that is, combining conservation with other uses.
4. Monitoring the state of the environment – using organisms to monitor pollution and to indicate the quality of the environment.
5. Monitoring as a way of advancing knowledge about the dynamics of ecosystems.
6. Monitoring of insect pests of agriculture and forestry so as to establish effective means of control of those pests.

These are all important objectives but they can not be achieved without the appropriate funding, staffing and logistical support (see comment on monitoring water quality in Britain in Chapter 12). Nor can they be achieved without enforcement of statutory requirements, at least where such statutory requirements apply.

A considerable challenge for future monitoring has arisen from research on genetically manipulated organisms (GMOs). There seem to be many unknowns about the implications of releasing GMOs and the possible long-term effects on ecosystems seem not to have been considered in depth. There is therefore a clear and urgent need for development of rigorous biological monitoring techniques as well as some careful thought being given to the logistics of such monitoring programmes. Who would do the monitoring and how would the data be stored and communicated? The potential applications of GMOs and the competitive research now taking place will not make it any easier to ensure that effective and rigorous monitoring does take place. There is, however, a potentially very important contribution yet to be made by ecologists if release of GMOs becomes more common.

Terms, concepts and aims

Definitions and questions about monitoring

Throughout this book, the terms monitoring, surveillance and census are not always used in a rigorous fashion and the aim here is not to undertake academic discussions about definitions. However, because these terms have been defined in various ways it would seem useful to offer a distinction between the terms as used in this book. Recording and also mapping, surveys, and sampling are all methods of data collection which provide a basis for the systematic measurement of variables and processes over time. Clark (1986) gives some good examples of the methods and applications for ecological monitoring.

The term census generally refers to population counts which in turn can be used in monitoring programmes. Surveillance is the systematic measurement of variables and processes over time, the aim being to establish a series of data in time. Monitoring is also the systematic measurement of variables and processes over time but assumes that there is a specific reason for that collection of data such as ensuring standards are being met. In a report of the Study of Critical Environmental Problems (SCEP 1970) entitled *Man's Impact on the Global Environment* there is a similar definition of monitoring which is:

systematic observations of parameters related to a specific problem, designed to provide information on the characteristics of the problem and their changes with time.

Differences between the different levels of monitoring, that is environmental, biological and ecological, may not always be rigorously adhered to.

Environmental monitoring covers a wide range of activites and a good, recent example of those activities can be seen in Craig's (1989) symposium volume on environmental monitoring in New Zealand. Monitoring climatic variables and processes is one very familiar example of environmental monitoring. That is, the systematic recording of soil and air temperatures, humidity, pressure and many other variables as well as processes is undertaken by meterological organizations throughout the world to provide a basis for a number of purposes, including predictive modelling. In addition to modelling, the monitoring of climatic variables and processes has played an important role in assessments of human impacts on climate, for instance monitoring temperatures has provided some of the evidence for global warming (Fig. 1.4).

A little reading will soon demonstrate that different authors have attributed different meanings to the term biological monitoring (as opposed to ecological monitoring) but one common definition in use, especially in the USA, is as follows:

biological monitoring is the regular, systematic use of organisms to determine environmental quality (Cairns 1979).

It was Cairns who played a major role in establishing standards and techniques for biological monitoring of water quality in the USA and indeed one important book (Cairns, Dickson & Maki 1978) changed the whole nature of biological monitoring in the USA because it connected fate and transport of pollutants to the biological effects. Some of the techniques developed and used by Cairns and others are described in later sections (see, for example, Chapter 12). Here we are using the term biological monitoring in a wider sense and not only for analysing the state of the environment by way of individuals organisms, populations and species. We are, for example, also concerned with monitoring populations and communities in order to understand ecological processes which may occur over long periods of time.

Rather than being concerned with definitions, it would seem more useful to think about the objectives and methods of monitoring. Even more important is the interpretation and practical application of the information which may be provided by a monitoring or surveillance programme. Two sets of questions can usefully be asked when considering any monitoring programme, one set of questions concerns the data collection and the second set is concerned with analysis and interpretation. Both sets of questions set the theme of this book.

Data collection.
1. Who is undertaking or supporting the programme?
2. What are the aims and objectives of the monitoring programme?

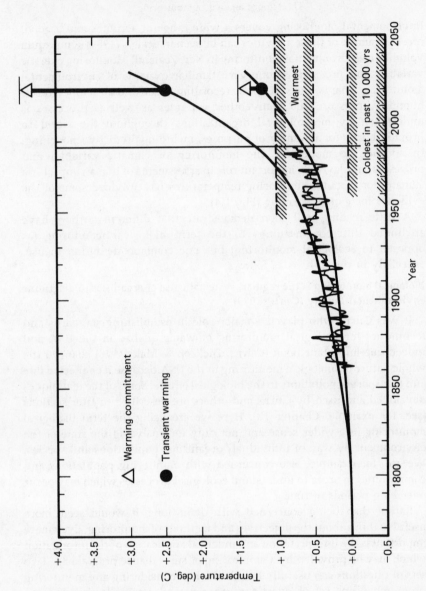

Fig. 1.4. Global mean temperature changes. Data from T.M.L. Wigley and redrawn after Holdgate (1991).

3. What methods are being used to systematically measure and record the variables and processes over time and are they the most appropriate methods?
4. What is the time-scale for the monitoring and what is the frequency of data collection?
5. What variables and processes have been chosen and are they the best variables for the objective of the programme?

Data analysis, presentation and interpretation.
1. What methods have been used for analysis of the data?
2. How are the data presented and are they presented in the most informative manner, bearing in mind the reader or audience?
3. Have the data been usefully and fully interpreted?

These seem to be obvious and simple questions but I suggest that they may encourage an enthusiastic and enquiring approach to a study of biological and ecological monitoring.

Aims of the book

The aim is to provide not only an introduction to the subject of biological and ecological monitoring but also to provide a basis for practical applications of monitoring from the most basic to the more complex. The scope is intended to be broad but with non-technical details about applications of monitoring. Thus it is hoped that students, especially during field courses, research students, as well as environmental managers, may find the selection of examples to be informative and thought provoking.

Some aspects of biological monitoring are not described here and these include the following: medical and epidemiological aspects of biological monitoring, monitoring of soil biology, biochemical aspects and microbiological monitoring. Biotelemetry and the study of animal behaviour (migration, home range studies and dispersion) is a vast subject beyond the scope of this book and passing reference only is made to the use of radio telemetry and other remote sensing techniques for the study of animal behaviour. The use of telemetry for studying and monitoring animal physiology is also not discussed in this book. Monitoring some aspects of pollution is included but not ecotoxicology nor monitoring for radionuclides. Biological monitoring in marine ecosystems is not included in detail.

Sources of information on soil and microbial monitoring include the *Canadian Journal of Microbiology*, and Worf (1980). The *Journal of*

Environmental Radioactivity includes articles on monitoring for radio-nuclides and a paper by Bada *et al.* (1990) provides an interesting introduction to monitoring environmental isotopic parameters by way of mammalian teeth. The journal *Biotelemetry* is an excellent source of information on the use of remote sensing for the study of animal ecology and behaviour. Pollution monitoring, especially biological pollution monitoring, is a topic very well addressed in the journal *Environmental Monitoring and Assessment* and there are special issues of that journal which report proceedings of symposia on bioindicators and monitoring (see, for example, Maher & Norris 1990). Ramade's book (*Ecotoxicology*) first published in 1977 is a useful text on pollution and monitoring. An excellent treatment of biological monitoring of marine ecosystems can be found in a number of texts including Kinne & Bulnheim (1980), Vernberg *et al.* (1981), Flemmer, Duke & Mayer (1986) and Wolfe *et al.* (1987). The Journal of the Estuarine Research Federation (*Estuaries*) has produced a dedicated issue on long-term biological data sets and monitoring (Champ *et al.* 1987).

Structure and contents

In the next two chapters of Part A, we look at the first two questions asked about monitoring: who does the monitoring and what are the objectives? The contents of Part B introduce some underlying ecology and ecological methods used in monitoring and then go on to consider various variables and processes used in monitoring.

Assessments of analysis, presentation and interpretation of data is the aim of Part C and this aim has been achieved by using a selection of topics and case studies. The choice of these topics and case histories has been determined by the material covered in the first two sections and in part by a desire to show good applications of biological monitoring but also to show areas where biological monitoring is in much need of better application. It was logical therefore to commence with a discussion about planning a monitoring programme, especially with regard to the importance of a conceptual plan or framework. The six topics which follow can be divided into two groups: the first includes topics in which biological monitoring is widely undertaken, the second group includes topics where there is much room for development of biological monitoring techniques and programmes.

Monitoring bird populations has been undertaken in many parts of the world by many organizations and by many individuals. There are those who would believe that birds are the ultimate indicators of the state of the environment and so for this reason the census, surveillance and monitoring of birds has been considered in some detail. Similarly, monitoring freshwater has initiated many kinds of monitoring programmes throughout the world and this allows us to assess many of the variables considered in Part B.

The remaining four sections (insularization, land use and landscapes, Environmental Impact Assessments, species monitoring and conservation) include examples of poorly planned monitoring and also instances where there is a need for new initiatives in biological monitoring. One of the most important threats to wildlife is habitat fragmentation yet there is little monitoring of the effects of fragmentation. Land use has been monitored for many years but there is a need for greater uniformity of the methods adopted. Environmental assessment and impact analysis is now common-place in many industrialized countries but monitoring is not often seen as an integral part of the Environmental Impact Assessment process. There are occasions when species monitoring, as opposed to community or ecosystem monitoring, can easily be justified but there is much need for more support of species monitoring programmes, especially when the rates of extinction are considered to be far greater than ever before witnessed (WCED 1987).

There are three appendixes. The first is an explanation of a simple dendrogram construction. The second gives addresses of most organizations mentioned in the text and the third appendix is an introduction to keys and literature for identification of taxonomic groups most commonly used in biological and ecological monitoring programmes.

References

Bada, J. L., Peterson, R. O., Schimmelmann, A. & Hedges, R. E. M. (1990). Moose teeth as monitors of environmental isotopic parameters. *Oecologia*, **82**, 102–6.
Benninghoff, W. S. & Bonner, W. N. (1985). *Man's Impact on the Antarctic Environment: A Procedure for Evaluating Impacts from Scientific and Logistic Activities*. Cambridge, SCAR and the Scott-Polar Institute.
Bonner, W. N. (1984). Conservation and the Antarctic. In *Antarctic Ecology*, ed. R. M. Laws, Vol. 2, pp. 821–50, London, Academic Press.
Cairns, J. (1979). Biological monitoring – concept and scope. In *Environmental Biomonitoring, Assessment, Prediction and Management – Certain Case Studies and Related Quantitative Issues*, ed. J. Cairns, G. P. Patil & W. E. Waters, pp. 3–20. International Cooperative Publishing House, Maryland.
Cairns, J., Dickson, K. L. & Maki, A. (1978). *Estimating the Hazard of Chemical Substances to Aquatic Life*. Special Technical Publication 657. American Society for Testing and Materials, Philadelphia, PA.
Champ, M. A., Wolfe, D. A., Flemer, D. A. & Mearns, A. J. (1987). Long-term biological data sets: their role in research, monitoring and management of estuarine and coastal marine systems. *Estuaries*, **10**, 181–93.

Clark, R. (ed.) (1986). *The Handbook of Ecological Monitoring*. Oxford, Clarendon Press.

Craig, B. (1989). *Proceedings of a Symposium on Environmental Monitoring in New Zealand with Emphasis on Protected Natural Areas*. Wellington, Department of Conservation.

Davis, M. B. (1989). Retrospective studies. In *Long-term Studies in Ecology: Approaches and Alternatives*, ed. G. E. Likens, pp. 71–89. New York, Berlin, Springer Verlag.

Engstrom, D. R., Swain, E. B. & Kingston, J. C. (1985). A palaeolimnological record of human disturbance from Harvey's Lake, Vermont: geochemistry, pigments and diatoms. *Freshwater Biology*, 15, 261–88.

Flemmer, D. A., Duke, T. W. & Mayer, F. L. (1986). Integration of monitoring and research in coastal waters: issues for consideration from a regulatory point of view. In *Oceans '86 Conference Record, Vol. 3, Monitoring Strategies Symposium*, pp. 980–92. Marine Technology Society, Washington, D.C.

Franklin, J. F., Bledsoe, C. S. & Callahan, J. T. (1990). Contributions of the long-term ecological research program. *BioScience*, 40, 509–23.

Gruber, D. & Diamond, J. (1988). *Automated Biomonitoring: Living Sensors as Environmental Monitors*. Chichester, Ellis Horwood.

Hardy, A. C. (1956). *The Open Sea, Its Natural History: The World of Plankton*, London, Collins.

Hassell, M. P. (1989). Workshop on Environmental Priorities. *Biologist*, 36, 275–80.

Holdgate, M. (1991). Conservation in a world context. In *The Scientific Management of Temperature Communities for Conservation*, pp. 1–26, ed. I. F. Spellerberg, F. B. Goldsmith & M. G. Morris, Oxford, Blackwell Scientific Publications.

Jeffers, J. N. R. (1989). Environmental monitoring. *Biologist*, 36, 171.

Kinne, O. & Bulnheim, H.-P. (1980). Protection of Life in the Sea. 14th European Marine Biology Symposium. *Helgolander Meeresuntersuchungen*, 33, 1–4. Biologische Anstalt Helgoland, Hamburg.

Likens, G. E. (1983). A priority for ecological research. *Bulletin of the Ecological Society in America*, 64, 234–43.

Likens, G. E. (1989). *Long-term Studies in Ecology: Approaches and Alternatives*. New York, Berlin, Springer-Verlag.

Maher, W. A. & Norris, R. H. (1990). International Symposium on Biomonitoring of the State of the Environment. *Environmental Monitoring and Assessment*, Special Issue, 14 (2–3).

MARC (1985). *Historical Monitoring*. Monitoring and Assessment Research Centre, University of London.

NERC (1989). *Proposed construction of a crushed rock airstrip at Rothera point, Adelaide Island, British Antarctic Territory*. Final comprehensive environmental valuation, BAS, NERC.

Orio, A. A. (1989). Modern chemical technologies for assessment and solution of environmental problems. In *Changing the Global Environment, Perspective on Human Involvement*, ed. D. B. Botkin, M. F. Caswell, J. E. Estes & A. A. Orio, pp. 169–84, London, New York, Academic Press.

Palmer, M. E. (1986). A survey of rare plant monitoring: programs, regions and species priority. *Natural Areas Journal*, 6, 27–42.

Parker, B. C. & Howard, R. V. (1977). The first environmental impact monitoring and assessment in Antarctica. The Dry Valley drilling project. *Biological Conservation*, 12, 163–77.

Ramade, F. (1977). *Ecotoxicology*. Chichester, John Wiley, English translation by L. J. M. Hodgson.

Ratcliffe, D. (1980). *The Peregrine Falcon*, Calton, T. & A. D. Poyser.

SCEP (1970). *Man's Impact of the Global Environment*. Report of the Study of Critical Environmental Problems (SCEP), Cambridge, MS., MIT Press.

Sladen, W. J. L., Menzie, C. M. & Reichel, W. L. (1966). DDT residues in Adelie penguins and a crabeater seal in the Antarctic. *Nature*, **210**, 670–3.

Strayer, D., Glitzenstein, J. S., Jones, C. G., Kolasa, J., Likens, G. E., McDonnell, M. J., Parker, G. G. & Pickett, S. T. A. (1986). *Long-term Ecological Studies: an Illustrated Account of their Design, Operation, and Importance to Ecology*. Occasional Publication of the Institute of Ecosystem Studies, No. 2, The N.Y. Botanical Garden, Millbrook, New York.

Tubbs, C. (1986). *The New Forest*. London, Collins.

Vernberg, F. J., Calabrese, A., Thurberg, F. P. & Vernberg, W. B. (1981). *Biological Monitoring of Marine Pollutants*. London, New York, Academic Press.

WCED (1987). *Our Common Future*. Oxford, Oxford University Press.

Wolfe, D. A. & O'Connor, J. S. (1986). Some limitations of indicators and their place in monitoring schemes. In *Oceans '86 Conference Record, Vol. 3, Monitoring Strategies Symposium*, pp. 878–84, Marine Technology Society, Washington, D.C.

Wolfe, D. A., Champ, M. A., Flemer, D. A. & Mearns, A. J. (1987). Long-term biological data sets: their role in research, monitoring, and coastal marine systems. *Estuaries*, **10**, 181–93.

Worf (1980). *Biological Monitoring for Environmental Effects*. Mass., Lexington Books.

2
World programmes and monitoring organizations

An historical perspective

LOOKING BACK through episodes of man's impact on the environment, it is difficult to know where to commence an introduction to national and international environmental monitoring activities and organizations. Many historians could possibly identify examples of environmental monitoring which took place many centuries ago but which can not be considered here for reasons of space. Significant events in the history of the exploitation of marine populations seem a useful place to start; it was in 1946 that an International Fisheries Convention was signed and the International Whaling Commission was established (Table 2.1). The need for marine biological monitoring at that time was evident but any practical use of monitoring data in the exploitation of marine populations was not to take place until many years later.

The year 1948 was a very important year for wildlife conservation and monitoring of endangered plants and animals. In that year the International Union for the Protection of Nature (IUPN) was established at a conference at Fontainebleau, convened through the initiative of Sir Julian Huxley during the time when he was Director-General of UNESCO. Later, in 1957 that organization was renamed the International Union for the Conservation of Nature and Natural Resources (IUCN).

Commencing in the early 1960s, a number of very important publications were to highlight the importance of environmental and biological monitoring. First, *Silent Spring* was published (Carson 1962) and despite criticisms about the emotive and less than scientific approach used in *Silent Spring*, that book prompted much concern about effects of pesticides, particularly the effects of pesticides on non-target organisms as a result of very widespread use of agriculture chemicals in the USA. Fifteen years on,

26

Table 2.1. *Some events and publications of significance to the development of environmental and biological monitoring*

1946	International Convention for the Regulation of Whaling establishes the International Whaling Commission.
1948	UN Charter.
	International Union for the Protection of Nature (IUPN) established.
1955	The Wenner Gren Conference on Man's Role in Changing the Face of the Earth, Wenner Gren Foundation, Princeton, New Jersey, USA.
1956	*Man's Role in Changing the Face of the Earth* (Thomas 1955) published.
1957	The IUPN becomes the International Union for the Conservation of Nature and Natural Resources (IUCN).
1958	Law of the Sea. The first UN Conference on the Law of the Sea approves draft conventions.
1959	Antarctic Treaty.
	Economic and Social Council of the UN adopts resolution to publish a register of national parks and equivalent reserves of the world.
1961	Establishment of World Wildlife Fund (World Wide Fund for Nature).
1962	*Silent Spring* (Carson 1962) published.
1964	International Council of Scientific Unions (ICSU) established the International Biological Programme (IBP).
1966	IUCN Red Data Books first published.
1968	UNESCO 'Biosphere' Conference.
1969	Friends of the Earth (FOE) founded.
1970	The US National Environmental Policy Act (NEPA) requires preparation of Environmental Impact Assessments.
1971	Man and the Biosphere (MAB) Programme of UNESCO launched.
	Greenpeace International founded.
1972	UN Stockholm Conference on the Human Environment.
	Concept of a Global Monitoring System (GEMS) endorsed by the Stockholm Conference.
	United Nations Environment (UNEP) Programme established.
	'Blueprint for Survival' sponsored by the journal Ecologist. *Limits to Growth* (Meadows *et al.* 1972) published.
1974	UNEP Regional Seas Programme established.
1975	Convention on International Trade in Endangered Species of Wild Fauna and Flora (CITES).
	The Kenya Rangeland Ecological Monitoring Unit (KREMU) established as a result of collaboration between Kenya and the Canadian International Development Agency.
1976	The Scientific Committee on Problems of the Environment (SCOPE) reports to the International Council of Scientific Unions (ICSU) on global trends in the biosphere most urgently requiring international and interdisciplinary scientific effort.
1977	UN Conference on desertification.
1979	World Climate Conference organized by the World Meteorological organization recognizes the 'greenhouse effect'.
1980	World Conservation Strategy (IUCN) launched.
	IUCN Conservation Monitoring Centre (now the World Conservation

Table 2.1. (*cont.*)

Monitoring Centre) established.

The Global 2000 Report to President Carter.

1982   UN Nairobi Conference, ten years after the Stockholm Conference.
World Charter for Nature adopted by UN.
The World Environment 1972–1982 is published (Holdgate, Kassas &
White, 1982).
International Union of Biological Sciences initiated the Bioindicators
Programme.

1983   UN General Assembly calls for establishment of an independent
commission: The World Commission on Environment and Development.

1984   *The Resourceful Earth. A Response to Global 2000* (Simon & Kahn 1984)
is published.
First of a series of symposia on monitoring programmes organised by
the International Union of Biological Sciences.

1985   European Community Environmental Impact Assessment Directive
adopted.
First Cary Conference: status and future of ecosystem science was
discussed.
Conference held in Venice on 'Man's Role in Changing the Global
Environment' (Orio & Botkin 1986).

1987   *Our Common Future* (WCED) is published.
Second Cary Conference (New York): endorsement of the need for long-
term sustained ecological research.

1989   The 'Green Summit' in Paris of seven industrial countries; final
communique says 'urgent need to safeguard the environment'.
IUCN publishes *From Strategy to Action*, a response to *Our Common
Future*.

1990   *BioScience* (July/August 1990, Vol. 40, No. 7) publishes three major
articles on long-term ecological research. See Franklin, Bledsoe &
Callahan (1990).

environmental concerns were an incentive for President Carter to call for
an environmental report on the future. The report *Global 2000*, published
in 1980 and prepared by the Department of State and the Council on
Environmental Quality, is a gloomy projection of what the world's
resources and the quality of life will be like in the next century (for a
review see Carter 1980). Much critical appraisal has been directed at the
report *Global 2000*, none more so than that in the publication *The
Resourceful Earth* (Simon & Kahn 1984). Although the need for biological
monitoring is argued indirectly in both these publications, we can not here
explore the arguments for reasons of space and therefore we need to focus
on publications and events more central to biological monitoring.

In the same year as the report *Global 2000* was published, another milestone report appeared, *The World Conservation Strategy* (IUCN 1980). Prepared by the IUCN, UNEP and WWF, this report focused on three objectives:

1. maintenance of ecological processes and life support systems;
2. preservation of genetic diversity and the conservation of wild species;
3. to ensure the sustainable utilization of species and ecosystems and to use all our natural resources carefully giving due consideration to the needs of future generations.

A successor volume to *The World Conservation Strategy* is currently being prepared, and should be published in 1991. Seven years after the publication of *The World Conservation Strategy*, the report of the World Commission on Environment and Development, *Our Common Future* or 'Bruntland Report' (WCED 1987), was heralded as the most important document of the decade on the future of the world. The 'Brundtland Report' re-examined the critical environment issues and development problems with the aim of formulating realistic proposals to solve the issues and problems but with sustainable development. The 'Brundtland Report' also highlighted the need for environmental monitoring in its many forms:

We need a kind of new earth/space monitoring system. I think that it goes further than simply an earth environmental system. It's a combined earth/space monitoring system, a new agency that would have the resources to be able to monitor, report, and recommend in a very systematic way on the earth/space interaction that is so fundamental to a total ecological view of the biosphere.

Maxwell Cohen, University of Ottawa, *Our Common Future*, p. 275
(WCED 1987).

Mention of human population growth and distribution can hardly be avoided, and quite rightly so, when discussing environmental issues and diminishing biological resources. Statistics on and monitoring of world population trends, health, agriculture, energy, fisheries and forests is undertaken by the United Nations Statistical Office (UNSO), the Food and Agriculture Organization of the UN (FAO) and the World Health Organization (WHO). The most important international event which was to act as a stimulus for global environmental monitoring was the 1972 United Nations Conference on the Human Environment (popularly called the 'Stockholm Conference') held at Stockholm in Sweden.

Following the 'Stockholm Conference', The United Nations Environment Programme (UNEP) was established. World-wide research on the environment as well as monitoring and information exchange were considered to be essential and an action plan agreed at Stockholm had three

major components: environmental assessment, environmental management and supporting measures. Ten years on from the Stockholm
Conference, The United Nations adopted the World Charter for Nature on
28 October 1982 and paragraph 19 in that Charter was confirmation of the
world-wide acceptance of a need for biological and ecological monitoring:

> The status of natural processes, ecosystems and species should be closely
> monitored to enable early detection of degradation or threat, ensure timely
> intervention and facilitate the evaluation of conservation policies and
> methods.   Para. 19, World Charter for Nature.

I am in no doubt that the next decade and the first few years of the
second millennium will see some very exciting advances in monitoring on
a global scale. Indications of those very exciting developments come from
the USA in the form of an Earth Observing System (EOS), a programme
which is related to a planned space station. The National Aeronautics and
Space Administration (NASA) plans to establish EOS in the next decade to
provide information on global processes, especially those biochemical,
biological, geophysical and socio-economic processes which change the
face of the earth. The success of such highly sophisticated and complex
programmes will, however, be dependent on a more detailed understanding of long-term ecological processes.

International monitoring organizations

In the space available it is not possible to do justice to all monitoring
organizations and therefore those mentioned here and in Chapter 3 are but
a selection. The United Nations Conference on the Human Environment
was a landmark in the development of world environmental monitoring
initiatives. Credit is due, however, to the many previous efforts which set
about to establish global monitoring such as that directed towards a
network of major monitoring stations in natural areas throughout the
world (MIT 1970). Many of these efforts are usefully outlined in a directory
published by the Smithsonian Institution (1970). The Special Committee on
Problems of the Environment (SCOPE) and other components of the
International Council of Scientific Unions were also to play an important
part in bringing these initiatives to the attention of the UN Conference.

UNEP and GEMS

The concept of a global environmental monitoring system was endorsed by
the UN Conference on the Human Environment in 1972, then later in 1975,
UNEP embarked on the most ambitious programme to monitor the quality
of air, water and food on a global scale (Holdgate & White 1977). The

Global Environmental Monitoring System (with the felicitous acronym GEMS) was established (with a Programme Activity Centre (PAC) at the UNEP Headquarters at Nairobi, Kenya) to acquire data for the better management of the environment. As a first step, the expertise of the WHO was enlisted because of that organization's experience in air monitoring. The first, and demanding, task to be tackled by GEMS was that of coordination, collection and dissemination of information from environmental monitoring programmes, particularly at the international level and through the services of The International Environmental Information Network (INFOTERRA). This information network was formerly known as the International Referral System for Sources of the Environment. The task of assembling and dissemination of data on a global scale seemed then, and still seems, extremely daunting despite the great improvements in methods of data collection and communication which have taken place over the last few years. GEMS has responded well to these demanding tasks and there seems no doubt that it is now the foremost world environmental monitoring agency.

The Governing Council of UNEP requires annual reports and comprehensive five-yearly reports on the state of the environment, a requirement laid down by the UN general assembly, namely to 'keep under review the world environmental situation in order to ensure that emerging environmental problems of wide international significance receive appropriate and adequate consideration'. The broad scope of topics in the Annual State of the Environment reports published during the first few years following the Stockholm Conference was impressive (Table 2.2) and showed that a truly international monitoring strategy had been achieved. For example, global air monitoring uses data from more than 60 countries and monitoring water quality takes place at 344 monitoring sites in 42 countries.

By 1982, monitoring activities supported by UNEP were firmly established under the auspices of GEMS and included the following.

1. Climate-related monitoring.
2. Monitoring of long-range transport of pollutants.
3. Health-related monitoring.
4. Ocean monitoring.
5. Terrestrial renewable-resources monitoring.

The principal activity of GEMS is environmental monitoring, an activity assisted by the expertise of specialized agencies; for instance, there are several agencies which contribute data on terrestrial renewable-resource monitoring (Table 2.3).

One example of 'terrestrial renewable-resources monitoring' is provided by the work carried out from 1980 to 1985 on the Sahelian pastoral

Table 2.2. *Topics treated in Annual State of the Environment Reports submitted to the Governing Council of* UNEP *from 1972 to 1982*

Subject area	Topics
The atmosphere	Climate changes and causes
	Effects of ozone depletion
Marine environment	Oceans
Freshwater environment	Quality of water resources
	Ground-water
Terrestrial environment	Land resources
	Raw materials
	Firewood
Food and agriculture	Food shortages, hunger, and degradation and losses of agricultural land
	Use of agricultural residues
	Pesticide resistance
Environment and health	Toxic substances and effects
	Heavy metals and health
	Cancer, Malaria, Schistosomiasis
	Chemicals in food chains
Energy	Energy conservation
Environmental pollution	Toxic substances
	Noise pollution
Man and the environment	Human stress and social tension
	Demography and populations
	Tourism
	Transport
	Environmental effects of military activity
	The child and the environment
Environmental management achievements	
	The approach to management
	Protection and improvement of the environment
	Environmental economics

From Holdgate *et al.* (1982).

ecosystems in Senegal (GEMS 1988), under the auspices of both FAO's 'Ecological Management of Arid and Semi-arid Rangelands' and UNEP's GEMS. The Sahel region of Africa is the semi-arid region to the south of the Sahara where there has been pronounced droughts over the last two decades. Several countries make up the Sahel region and these include the following: Burkina Faso, Chad, Djibouti, Ethiopia, Mali, Mauritania, Niger, Senegal, Somalia and the Sudan.

The GEMS programme had previously specified that 'monitoring activities will be undertaken and expanded following the recommend-

Table 2.3. *Examples of monitoring projects implemented by UNEP, specialized agencies of the UN and non-governmental organization*

Project title	Headquarters location	Date founded	Participants
GEMS – Global Environment Monitoring System	UNEP Nairobi	1975	Global
Biological monitoring pilot project	WHO Geneva	1978	10 countries
Tropical forests resources assessment	FAO Rome		76 countries
Tropical forest cover monitoring	FAO Rome		3 countries
Pastoral ecosystem monitiring in West Africa	LNERV/ISRA, Daker		
Desertification monitoring in Latin America	ONERN, Lima		
Soil degradation in North Africa and Middle East	FAO Rome		
Monitoring status of mammals	WCMC Cambridge	1980[a]	Global
Monitoring status of birds	ICBP Cambridge	1980[a]	Global
Trade in endangered species	WCMC Cambridge	1980[a]	Global
Coral reefs	WCMC Cambridge	1980[a]	Global
Parks and protected areas	WCMC Cambridge	1981	Global

From UNEP (1987).
[a] Date for commencement of UNEP–GEMS association.
LNEVR, Laboratoire National d'Elevage et de la Recherche Veterinaire; ISRA, Institute Senegalais de Researches Agricoles; ONERN, Officina Nacional Evaluacion Recursos Naturales; ICBP, International Council for Bird Preservation; WCMC, World Conservation Monitoring Centre.

ations of the group of government experts which will examine monitoring as a means of evaluating problems resulting from agricultural and land-use practices'. The main objectives therefore of the Sahelian programme included the provision of baseline data, the establishment of standard methods on an international level for the monitoring of the rangelands and to suggest actions to combat desertification (GEMS 1988).

A test area of 30 000 km² was selected for the Sahelian Pastoral Ecosystem monitoring project and monitoring was undertaken with satellite imagery, aerial photography, low-altitude flights, and ground control validation. Three main lines of activity developed from the project; satellite evaluation of green biomass on the range at the end of the rainy season, livestock census and evaluation of erosion, control of remotely sensed data via ground sampling. The data have made it possible to forecast the management needs for the nine months of the dry season and therefore to improve management of livestock in an area subject to drought and desertification.

As well as the GEMS programmes in the Sahel, the IUCN (with generous support from Nordic Countries) has, more recently, become involved in the challenging environmental problems and the desperate plight of millions of people in the Sahel. Following the establishment in 1987 of a Sahel Coordination Unit at IUCN Headquarters in Gland, one of the largest IUCN field programmes has been established in the Sahel with a budget of around SFr 6 million. The IUCN Sahel programme's objectives include:

1. to develop ways to manage living natural resources that better correspond to prevailing climatic conditions and which permit sustainable development,
2. to help preserve the biological diversity of the Sahel,
3. to monitor the changes taking place in the Sahel region.

A recent report prepared by IUCN and the Norwegian Agency for International Development (IUCN 1989a) has provided an assessment of the major issues affecting the Sahel region, a report which usefully combines studies on both the people and the environment as a basis for commencing plans which hopefully will lead to sustainable development.

Remote sensing by satellite imagery, a method which is now undertaken by GEMS, has greatly enlarged the scope of monitoring (see p. 258). In 1985, UNEP established a new element of GEMS in the form of the Global Resource Information Database (GRID) and the key to this was the successful launch of the SPOT-1 satellite aboard the European Ariana 16 rocket on 22 February 1986. The acronym SPOT refers to System Probatoire d'Observation de la Terre. GRID has two centres, the control facility in Nairobi with particular emphasis on Africa, and the Geneva centre which focuses on global and continental data sets. Advanced technology and global data sets via numerous organizations (including UNEP, UNESCO, FAO, WHO and the IUCN) enable GRID to assess and examine interactions between different environmental data sets (see Gwynne & Mooneyhan 1989 for further details).

The applications of the GRID system are many and we are not able to do justice to the system in this brief account. However, one recent application was in connection with monitoring and conservation of African Elephants leading to one conclusion that the off-take of ivory was more than the

Elephant populations could sustain and that many populations were in danger of extinction (UNEP/GEMS 1989). Following on from this, the recent action by the Convention for International Trade in Endangered Species (CITES) highlighted the serious damage inflicted on elephant populations by ivory poaching. The reports from CITES have, in turn, forced many nations to give urgent consideration to serious implications of sustaining the ivory industry (see Caughley, Dublin & Parker 1990 for an interesting account on the decline of the African Elephant). Monitoring Elephants and the ivory episode was just one small example of the potential of GRID, a potential which seems less than well appreciated and understood by the international community. Indeed, much can be learnt from the African experiences of monitoring; a wealth of monitoring which has been undertaken by international agencies as well as national centres of research such as the National Programme for Ecosystem Research based in Pretoria.

Many reports have now been published by UNEPS and GEMS but one 'landmark' publication came in 1987 when the UNEP Environmental Data Report was published. Prepared by the Monitoring and Assessment Research Centre (MARC) in cooperation with other agencies, this report summarizes the 'best environmental data that are currently available

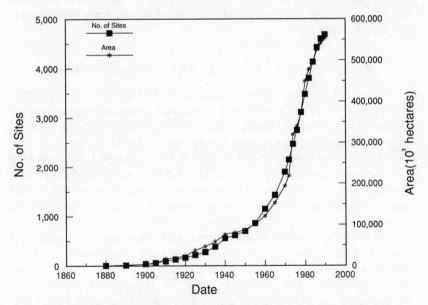

Fig. 2.1. Growth in the number and area of nationally protected areas (nature reserves, national parks, natural monuments and wildlife sanctuaries). Data kindly supplied by The World Conservation Monitoring Centre.

today'. The agency MARC (see below) maintains a computer database on environmental information based on the UNEP Environmental Data Report and this database will be updated as new information becomes available. A section on natural resources is included in the report along with eight other sections: environmental pollution, climate, populations/settlements, human health, transport/tourism, wastes and natural disasters. Concentrating mainly on land use statistics, the section on natural resources provides basic data with broad categories for monitoring changes in usage.

Included amongst those statistics are data (from the World Conservation Monitoring Centre, Cambridge) which give cumulative totals of major nationally protected areas (Fig. 2.1). This kind of information is not easy to collect and the number of protected areas shown in Fig. 2.1 is greater than previously quoted in the literature (see Chapter 13). These nationally protected areas are based on the IUCN management categories I–V and include (I) scientific reserves and nature reserves, (II) national parks, (III) natural monuments and natural landmarks, (IV) managed nature reserves and wildlife sanctuaries and (V) protected landscapes. Monitoring the area and number of protected areas as part of land use statistics is gratifying but the extent of protection afforded to wildlife by these areas is extremely varied (see Chapter 13). The effectiveness of protected areas as a basis for nature conservation has more recently been brought into question, especially because of the effects of isolation on some populations (Spellerberg 1991).

Monitoring and Assessment Research Centre (MARC)

MARC, an independent, international environmental research institute, was established in 1975 to assist international organizations engaged in environmental projects. Based at King's College, University of London, MARC is funded by UNEP, GEMS, WHO and its host University. It is an interdisciplinary research centre which has developed five research programmes: assessment of global pollution problems, environmental data base assembly, environmental management and pollution control, environmental health protection and the development of international links for education and training in developing countries.

The MARC's assessment and research on biological and ecological monitoring has been directed at pollution, both retrospectively and in relation to present day use of plants to monitor chemical emissions and other pollutants. Historical monitoring or retrospective studies are important for the establishment of base-levels which occurred during pre-industrial times and can be used to determine man-made changes in pollution concentrations. The material used for historical monitoring is especially interesting and includes aqueous sediments, ice and snow, peat, plant tissues, museum collections, herbarium specimens and human remains. One of the richest sources used for historical monitoring is the sedimentary record, providing data on pollution from both effluents and atmospheric deposition. The sediments of the Great Lakes of North America for example have probably been the source of more data for historical monitoring than anywhere else in the World (MARC 1985). The combination of suitable sedimentary features for accumulating pollutants and the industry and agriculture surrounding the Great Lakes has resulted in marked concentrations of metals in the more recent sediments. For instance the inputs of Zn and Cu which have increased markedly this century can be attributed to the burning of coal and in some instances to the electroplating industry.

Materials of a biological origin which have been used in historical monitoring are particularly relevant to our discussions here and in later sections. MARC lists an interesting range of such material including herbarium specimens of mosses and lichens, zoological specimens of fish, bird feathers and eggs, animal hair, horns and teeth. The use of mosses and lichens as indicator species (see Chapter 5) for the detection of airborne pollutants is particularly important and past records could be established using herbarium collections. The use of bird feathers has been shown to be very effective in the historical monitoring of Hg in marine and terrestrial environments (MARC 1985).

Field Studies Council Research Centre

The Field Studies Council Research Centre (FSCRC) was established as the Oil Pollution Research Unit (OPRU) at the Orielton Field centre in Wales in 1967 as the result of an initiative following a spill of over 250 tonnes of crude oil in Milford Haven (southwest Wales). Since 1973 the FSCRC has been involved in surveys of North Sea oil-fields and from 1976 it was involved in many coastal survey and monitoring projects throughout the world (Dicks 1987). For example, typical projects included the establishment of sampling points for long-term monitoring programmes around an oil terminal in the Gulf of Suez, and rocky shore monitoring to assess ecological impacts of an oil terminal at Sullome Voe (Shetland Islands).

In 1987, the Field Studies Council took the decision to base all contract-funded research in South Wales around the FSCRC and thus offer a research and advisory service in the biological, chemical, physical and environmental sciences. About 20 scientists now make up the FSCRC.

International monitoring and surveillance of wildlife and natural resources

IUCN and the World Conservation Monitoring Centre

Biological monitoring, especially the monitoring of biological diversity, is one of the recommended actions suggested in that IUCN response to the 'Brundtland Report' (WCED 1987):

Recommended Action: Biological Diversity

Enhance monitoring of biological diversity. IUCN and other organizations like UNESCO's Man and the Biosphere Programme should establish a global system of 'Biological Diversity Monitoring Stations' against which environmental perturbations can be evaluated: these stations will study, among other parameters, the frequency and condition of certain key indicator species.

From IUCN (1989*b*).

The International Union for Conservation of Nature and Natural Resources (IUCN) is the world's largest conservation organization and was founded in 1948 at Fontainebleau, France. The Headquarters is in Switzerland and the main components of the IUCN include the General Assembly (which meets every three years) and currently six commissions, the status of which was discussed at the IUCN General Assembly in Perth in 1990. The Standing Commissions include: 1, Ecology, 2, Environmental Education and Training, 3, Sustainable Development, 4, Environmental Policy, Law and Administration, 5, National Parks and Protected Areas, and 6, Species Survival. The newsletter of the Species Survival Commission (*Species*) is a particularly informative publication sponsored by the World Wide Fund for Nature, UNEP and the IUCN.

The IUCN thematic programmes (Fig. 2.2) include tropical forests, wetlands, marine ecosystems, plants, the Sahel, Antarctica, populations and sustainable development, and women in conservation. This range of themes enables IUCN and its member organizations to develop sound policies and programmes for conservation of biological diversity and sustainable development of natural resources throughout the world.

The IUCN created the World Conservation Monitoring Centre (previously the Conservation Monitoring Centre, CMC) in 1980 to provide an improved environmental information service for use by conservation bodies, development agencies, governments, industry, scientists and the media. The WCMC is directed by an independent Board, directors of which are appointed by the IUCN, WWF and UNEP. It has developed integrated databases covering several themes which include: data on plant and animal

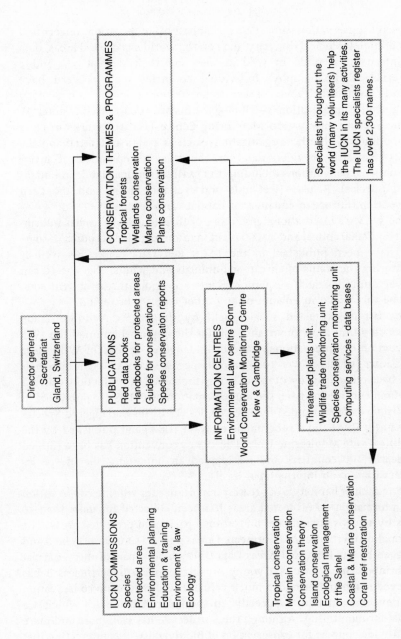

Fig. 2.2. Generalized structure of the International Union for the Conservation of Nature and Natural Resources. Arrows indicate direction of flow of information. See Fig. 16.5 for information about the World Conservation Monitoring Centre.

Director general
Secretariat
Gland, Switzerland

CONSERVATION THEMES & PROGRAMMES
Tropical forests
Wetlands conservation
Marine conservation
Plants conservation

Specialists throughout the world (many volunteers) help the IUCN in its many activities. The IUCN specialists register has over 2,300 names.

PUBLICATIONS
Red data books
Handbooks for protected areas
Guides for conservation
Species conservation reports

INFORMATION CENTRES
Environmental Law centre Bonn
World Conservation Monitoring Centre
Kew & Cambridge

Threatened plants unit.
Wildlife trade monitoring unit
Species conservation monitoring unit
Computing services - data bases

IUCN COMMISSIONS
Species
Protected area
Environmental planning
Education & training
Environment & law
Ecology

Tropical conservation
Mountain conservation
Conservation theory
Island conservation
Ecological management
of the Sahel
Coastal & Marine conservation
Coral reef restoration

species of conservation value, data on habitats of conservation concern and sites of high biological diversity, data on the world's protected areas, data on international trade in wild species and their derivatives, and a conservation bibliography. Individual countries such as Peru have recently created their own conservation data centres and Britain has long had a similar organization, the Biological Records Centre (see Chapter 3).

The World Conservation Monitoring Centre (WCMC) works in close collaboration with other organizations such as GEMS and GRID and the WCMC has also provided data for the effective operation of major international conventions including CITES (International Trade in Endangered Species), Ramsar (wetlands and wetland birds), and the Bern Convention (European endangered habitats and species).

The WCMC's international inventory of threatened species has information on 17 000 animal and 30 000 plant taxa and detailed accounts of some taxa have been published in the IUCN Red Data Books. As well as monitoring the status of threatened animals and plants, the WCMC can answer questions such as 'which plants are under threat or will soon become threatened in Poland' or any other country or region.

The list of protected species held by the WCMC Wildlife Trade Monitoring Unit is an important basis for the implementation of the 1975 Convention on International trade in Endangered Species of Wild Flora and Fauna (CITES). Over 1000 animal species, many of them mammals have been recognized by CITES as being threatened by extinction. Current developments arising from the implementation of CITES are published in the *Traffic Bulletin*, funded by The Peoples' Trust for Endangered Species (TRAFFIC, USA) and a programme of WWF (USA) and published by the Wildlife Trade Monitoring Unit. The CITES convention has been ratified by nearly 100 countries and has played an important role in species conservation at an international level.

Protection of habitats is of greatest importance for wildlife conservation and information on protected areas has been collected for more than 20 years by IUCN's Commission on National parks and protected areas. The information on thousands of protected areas (Fig. 2.1) is currently held and managed by the Protected Areas Data Unit which is part of the WCMC at Cambridge. The WCMC is a young organization but nevertheless it has achieved much and provided much valuable information on conservation and development issues especially in the form of databases which are crucial for monitoring. Although the last decade has seen more and more world-wide support for conservation of biodiversity, it is interesting that only recently (August 1989) has the USA National Science Foundation given its support to a programme which includes an inventory of all living organisms.

International Whaling Commission

The International Whaling Commission (IWC) was established in 1946 primarily to support commercial whaling and that role, responsible for over-exploitation, continued unabated until the 1960s. The IWC can not realistically be called a biological monitoring organization but it has been a forum for debates about whale populations and has commissioned investigations into whale stocks. In 1972, following the Stockholm Conference, the IWC rejected a moratorium on commercial killing of Whales which had been supported by 53 nations. Since that time, attempts to implement a moratorium have had a chequered history: an indefinite Sperm-whaling moratorium in 1981, a general moratorium in 1986 and an end to commercial whaling in 1988. Two years later, however, Norway was arguing the case for commercial exploitation of the Minke Whale.

If ever there were a strong case for heeding advice then whaling could not be a better example. The diminishing whale catches which have occurred since the 1940s have demonstrated very clearly that there was over-exploitation of these mammals. Yet, in the absence of 'better monitoring' of Whale populations, the folly of whaling has continued until very recently, sometimes under the guise of 'scientific whaling'. It is ironic that in 1989, the most accurate estimate of Blue Whales ever undertaken has shown that populations of that species are lower than previously feared by conservation organizations (see Chapter 16).

The IWC Scientific Committee has, over the last few years, been actively engaged in the preparation of 'comprehensive assessments' of Whale stocks. For example, the first of the comprehensive assessments of North Pacific Gray Whales, Southern Hemisphere and North Atlantic Minke Whales were planned to be presented to the IWC in 1990. After that time, it may be possible to have more accurate data for monitoring Whale populations.

Sea Mammals Research Unit

The Sea Mammals Research Unit (SMRU) is part of the Marine Sciences Directorate of the UK's Natural Environment Research Council and was established in 1977 by the amalgamation of two groups, one working on Whales and one working on Seals. The Unit is based in Cambridge and is concerned with all aspects of the ecology of Seals and Whales but it has a special expertise in estimating population levels and monitoring changes in those levels.

WWF, FOE and Greenpeace

As well as the IUCN and the CMC, there are many other organizations actively involved in the surveillance and census of wildlife and natural resources. These include the World Wide Fund for Nature (WWF, previously the World Wildlife Fund, founded in 1961), the Friends of the Earth (FOE) founded in 1969, and Greenpeace which was founded in 1971. These three organizations have sponsored research for biological monitoring, commissioned monitoring reports and have taken an active participation in the international surveillance of wildlife and natural resources. The WWF is an international organization with headquarters in Switzerland and national groups in about 30 countries. Mainly a fund raising organization, WWF has sponsored international research on species and wildlife throughout the World. A recent WWF campaign, 'Biodiversity a Conservation Imperative', has promoted the work of the World Conservation Monitoring Centre by listing plant and animal species on the brink of extinction.

References

Carson, R. (1962). *Silent Spring*, Boston, Houghton Mifflin.

Carter, L. J. (1980). Global 2000 Report: vision of a gloomy world. *Science*, **209**, 575–6.

Caughley, G., Dublin, H. & Parker, I. (1990). Projected decline of the African elephant. *Biological Conservation*, **54**, 157–64.

Dicks (1987). The Field Studies Council Oil Pollution Research Unit – the 1980s and beyond, *Biological Journal of the Linnean Society*, **32**, 111–26.

Franklin, J. F., Bledsoe, C. S. & Callahan, J. T. (1990). An expanded network of scientists, sites, and programs can provide crucial comparative analyses. *BioScience*, **40**, 509–23.

GEMS (1988). *The Global Environment Monitoring System*, Sahel Series main report. AG:EP/SEN/001 Technical Report. UNEP/FAO, Rome.

Gwynne, M. D. & Mooneyhan, D. W. (1989). The Global Environment Monitoring System and the need for a global resource data base. In *Changing the Global Environment, Perspectives on Human Involvement*, ed. D. B. Botkin, M. F. Caswell, J. E. Estes & A. A. Orio, pp. 243–56. Boston, London, Academic Press.

Holdgate, M. W., Kassas, M., & White, G. F. (eds.) (1982). *The World Environment, 1972–1982: A Report by the United Nations Environment Programme*. Dublin, Tycooly Int. Pub.

Holdgate, M. W. & White, G. F. (eds.) (1977). *Environmental Issues*. SCOPE Report 10. London, New York, Wiley.

IUCN (1980). *World Conservation Strategy: Living Resource Conservation for Sustainable Development*. Gland, IUCN–UNEP–WWF.

IUCN(1989a). *The IUCN Sahel Studies*, IUCN Regional Office for Eastern Africa, Nairobi.

IUCN(1989*b*). *From Strategy to Action. The IUCN Response to the Report of the World Commission on Environment and Development*, Gland, IUCN.

MARC (1985). *Historical Monitoring, A Technical Report*. London, MARC.

Meadows, D. H., Meadows, D. L., Randers, J. & Behrens, W. W. (1972). *Limits to Growth*. Earth Island, London, Potomac Associates Edn.

MIT (1970). *Man's Impact on the Global Environment: Assessment and Recommendations for Action. A Report of the Study of Critical Environmental Problems (SCEP)*. Cambridge, Mass., MIT Press.

Orio, A. A. & Botkin, D. B. (1986). Man's Role in Changing the Global Environment. *The Science of the Total Environment*, **55**, 1–400 & **56**, 1–416.

Simon, J. L. & Kahn, H. (1984). *The Resourceful Earth: A Response to Global 2000*, Oxford, Basil Blackwell.

Smithsonian Institution (1970). *National and International Environmental Monitoring Activities – a Directory*. Cambridge, Mass., Smithsonian Institution Center for Short-lived Phenomena.

Spellerberg, I. F. (1991). A biogeographical basis of conservation. In *The Scientific Management of Temperate Communities for Conservation*, pp. 293–322, ed. I. F. Spellerberg, F. B. Goldsmith & M. J. Morris. Oxford, Blackwell Scientific Publications.

Thomas, W. L. (ed.) (1956). *Man's Role in Changing the Face of the Earth*. Chicago, University of Chicago Press.

UNEP (1987). *United Nations Environment Programme, Environmental Data Report*. Oxford, Basil Blackwell.

UNEP/GEMS (1989). *The African Elephant*, UNEP/GEMS Environment Library No. 3, UNEP, Nairobi.

WCED (1987). *Our Common Future*. Oxford, Oxford University Press.

3

Biological monitoring in the United States and in Europe

Introduction

It would not be practical to describe all national biological monitoring organizations throughout the world and therefore a selective account has to be adopted. The United States and Europe has been chosen for no other reason than because of the wide range of examples of monitoring organizations in those two parts of the world. They have not been chosen because of any special quality or lack of quality of the monitoring activities.

Monitoring in the United States

Environmental monitoring has been undertaken in the USA for many years: for example, hydrological monitoring by the US Geological Survey, and climatic studies by the National Oceanographic and Atmospheric Administration. The monitoring undertaken by the US National Oceanic and Atmospheric Administration consists of many programmes but is directed largely at contaminant levels in sediments and tissues, and on measures of associated biological effects in organisms (as opposed to the effects at the ecosystem level). One such programme is in Marine Monitoring and Prediction (MARMAP). A useful source of information on these monitoring activities may be found in the annual reports: Reports to Congress on Ocean Pollution, Monitoring and Research (US Department of Commerce, NOAA) and in a bibliography of NOSS-sponsored ocean pollution monitoring, published in 1990 (NOAA 1990).

Long-term ecological programmes and biological monitoring have also been undertaken or promoted in the USA for many years by several agencies (including the Environmental Protection Agency, the National Science Foundation, the US Fish and Wildlife Service, the Smithsonian

Institution and the non-governmental body, the Nature Conservancy) but only more recently has the value of long-term biological and ecological monitoring been given special emphasis, especially at scientific meetings such as the second Cary Conference which was held in 1987. Over 60 scientists from 11 nations met at that Conference and endorsed the need for sustained ecological research and agreed that such research is a necessary prerequisite for monitoring and interpreting ecological data (see Chapter 1).

The National Science Foundation (NSF), Division of Biotic Systems and Resources, currently administers long-term ecological research programmes throughout a national network of research sites which are fairly representative of ecosystems in North America (Brenneman 1989) and include examples of forest, tundra, agricultural and prairie ecosystems (Table 3.1). The initiative for these programmes, which embrace aspects of ecosystem monitoring and surveillance, came largely from a series of workshops and meetings about fundamental issues in ecology (Botkin 1980). This long-term ecological research programme, although a pilot programme, was supported because it was recognized that many ecological phenomena occur on scales much longer than that normally supported by traditional research, and because it was agreed that long-term trends in ecosystems were simply not being monitored (Brenneman 1989). It is hoped that data collected from the recording sites will address the following:

1. pattern and control of primary productivity,
2. dynamics of populations of organisms selected to represent trophic structure,
3. pattern and control of organic matter accumulation in surface layers and sediments,
4. patterns of inorganic inputs and movements of nutrients through soils, groundwater, and surface waters,
5. patterns and frequency of disturbances.

The Nature Conservancy (NC) was founded in 1951 in the USA and is a very successful private conservation organization. The NC operates a network of over 1000 nature reserves and administers many important monitoring programmes, especially with the aim of providing good data for management and conservation of plant and animal species. The NC has sponsored detailed assessments of monitoring on nature reserves with the objective of identifying the most important monitoring needs and appropriate cost-effective methods for monitoring. One example of an NC nature reserve monitoring programme which looked particularly at the effects of grazing is described in Chapter 16.

Table 3.1. *Long-term ecological research sites (LTERs). Examples of sites in North America*

Site name	Institutional affiliation	LTER research topics	Principal biome
H. J. Andrews Experimental Forest	Oregon State University U.S. Forest Service Pacific Northwest Research Station	Successional changes of composition, structure, and processes Nature of forest–stream interactions Population dynamics of forest stands Effects of nitrogen fixers on soils Patterns and rates of log decomposition Disturbance regimes in forest landscapes	Coniferous forest
Arctic LTER site	Marine Biological Lab. University of Alaska Univ. of Massachusetts Clarkson University Univ. of Minnesota Univ. of Cincinnati Univ. of Kansas	Movement of nutrients from land to stream to lake Changes due to anthropogenic influences Controls of ecological processes by nutrients and by predation	Arctic tundra, lakes, and streams
Central Plains experimental range (CPER) – Shortgrass Steppe	US Department of Agriculture Agricultural Research Service (ARS) Colorado State Univ.	Hydrologic cycle and primary production Microbial responses Plant succession Plant and animal population dynamics Plant community structure Organic matter aggregation or degradation Influence of erosion cycle on redistribution of matter, nutrients, and pedogenic process Influence of atmospheric gases, aerosols, and particulates on primary production and nutrient cycles	Grassland

Harvard Forest	Harvard University	Long-term climate change, disturbance history and vegetation dynamics Comparative ecosystem study of anthropogenic and natural disturbance Community, population and plant architectural response to disturbance Forest–atmosphere trace gas fluxes Ecophysiology and micrometeorology Organic matter accumulation, decomposition, and mineralization Element cycling, the root dynamics and forest microbiology	Temperate-deciduous coniferous forest
Konza Prairie Research Research Natural Area	Kansas State Univ.	Role of fire, grazing, and climate in a tallgrass prairie ecosystem	Tallgrass prairie
North Inlet (Hobcaw Baronyl)	Univ. of South Carolina Belle W. Baruch Institute for Marine Biology and Coastal Research	Patterns and control of primary production Dynamics of selected populations Organic accumulation Patterns of inorganic contributions Patterns of site disturbances	Coastal marine

From Brenneman (1989).

Environmental and biological monitoring in Europe

CORINE Biotopes Programme

The acronym CORINE stands for Co-ORdination of INformation on the Environment, an experimental programme of the Directorate-General for the Environment, Nuclear Safety and Civil Protection of the Commission of the European Communities (Schneider 1989). In 1982, a European mapping case-study entitled 'Biotopes of Significance for Nature Conservation' was undertaken and this led in 1985 to the Commission adopting the CORINE Biotopes Programme (Rhind *et al.* 1986). The prudent and pragmatic objectives for the first four years were as follows.

1. An inventory of biotopes of major importance for nature conservation in the community.
2. Collating and making consistent data on acid deposition, and in particular the establishment of a survey on emissions into the air.
3. The evaluation of natural resources in the southern part of the community, in particular in those regions which are eligible for support.
4. Work on the availability and comparability of data.

The first four years of CORINE are currently being reviewed and meanwhile the Commission has announced the creation of a European Environmental Agency and a European Network for monitoring environmental information. Not before time has the Community addressed the all-important issues of environmental monitoring but we have yet to see what form the data gathering will take.

The Natural Environment Research Council (NERC) has been involved with CORINE since its planning stages. Wyatt (pers. comm.) of the Environmental Information Centre (based at Monks Wood) has undertaken a pilot study to catalogue important nature conservation areas. This is part of the programme of work to establish biotopes databases, using sophisticated computer programmes (Moss 1990). Despite the use of sophisticated computing facilities, the problems so far encountered by CORINE are basic, such as the need for compatibility between data gathering systems and good communication between those involved. A conceptual framework for monitoring within the CORINE programme seems to have addressed most if not all these basic problems (see Chapter 10 for comments about a conceptual framework for monitoring).

There are a few national monitoring programmes in Europe, such as the Swedish Environmental Monitoring Programme (National Swedish Environmental Protection Board 1985). This programme monitors various environmental variables, including ecological variables, at a range of sites. Awareness of marine pollution problems are not new in Europe but cooperative, regional monitoring has taken many years to become established. Under the Helsinki Convention on the Pollution of the Marine Environment of the Baltic, marine biological monitoring was based on data gathered on species abundance, fish population densities, biomass and diversity. Changes in the phytoplankton and other communities have also been monitored at several sites but the value and reliability of these variables for monitoring have been questioned (Morris, Samiullah & Burton 1988). Some of these variables are difficult to monitor in the field and are difficult to equate with specific pollutants but of more importance has been the difficulty in obtaining complete comparability of results from cooperative programmes.

Biological and ecological monitoring in Britain

Monitoring surveillance and census of biological systems in Britain is undertaken or sponsored by both Government and non-governmental organizations (NGOs). For many of them, biological monitoring is not a major objective but nevertheless there seems to be a growing interest in the value of monitoring especially with regard to nature conservation. For example, the Countryside Commission undertakes habitat surveys and has recently sponsored a monitoring project directed at management of wildlife and habitats on demonstration farms (see Chapter 10). The Royal Society for the Protection of Birds, The Mammal Society, the British Herpetological Society, The Botanical Society of the British Isles (sponsored by the Nature Conservancy Council), the British Butterfly Conservation Society and the British Lichen Society are just some of the NGOs who have sponsored or supported monitoring, surveillance or census of wildlife and habitats. The British Lichen Society's lichen mapping scheme started in 1965 and is based on the distribution of species within 10 km grids. The lichen mapping scheme is proving to be a particularly valuable scheme, partly because lichens are good indicator species (see Chapter 5) and partly because of the well-organized manner in which the scheme is operated. An atlas of the brophytes is also soon to be produced for the British Isles.

Coordination of views and interests amongst non-governmental conservation organizations has been undertaken by Wildlife Link, which was established in 1980. This is a liaison organization for all the major voluntary conservation groups in the UK who are concerned with the protection of wildlife. Of the many reports sponsored by Wildlife Link, some have been directed at a record of losses in protected areas such as Sites of Special Scientific Interest.

Field Studies Council

The Field Studies Council (a company limited by guarantee) has the main aim of increasing environmental understanding for all. That aim is achieved by a rich range of courses at their field centres, by publications and by research. A national network of Field Studies Council sites has provided a useful network for various ecological monitoring programmes. The Field Studies Council has collaborated with other environmental and educational organizations such as the WATCH Trust for Environmental Education in order to promote countrywide projects. For example, the WATCH Acid Drops project (a project devised to allow children throughout the country to monitor rainfall and acidification) was organized by the WATCH Trust in association with the Field Studies Council (Thomson 1987).

Department of the Environment

In 1986, the Department of the Environment (DOE) updated the DOE publication of 1974 entitled The Monitoring of the Environment in the United Kingdom (HMSO 1986). Mainly concerned with monitoring pollution, environmental monitoring in the UK has also included landscape monitoring and monitoring of land use change (Table 3.2). A discussion about monitoring land use and landscapes follows in Chapter 14.

The decentralized aspect of environmental activities in the UK makes it difficult to maintain an overview of the various programmes. For this reason the DOE has established a detailed, computerized register of environmental monitoring schemes sponsored or carried out by central government. In 1986, the register included some 400 schemes covering mainly the DOE's air pollution monitoring schemes, some water monitoring schemes and programmes of radioactivity monitoring. In addition to pollution monitoring, it is planned to include schemes on fauna and flora, especially habitat quality and endangered species and extinctions.

Natural Environment Research Council

The Natural Environment Research Council (NERC) does not undertake major monitoring programmes *per se* but has an interest in a number of monitoring programmes, including monitoring the health of the oceans with detailed continuous plankton recording. The NERC has also been instrumental in the establishment of the Environmental Information Centre (including the Biological Records Centre). An intended new initiative which is currently under discussion between the NERC and the

Table 3.2. *Examples of UK Government implemented monitoring programmes*

1. National Survey of Air Pollution. Established 1961.
   Monitoring of smoke and SO_2.
2. Acid Deposition. Established 1986.
   Rainfall analysed for ions, SO_2, NO_2.
3. Airborne Lead Concentrations, 1984–87 (Steering Committee on
   Environmental Lead Monitoring).
   Monitoring lead in air.
4. River Quality Survey, 1958, 1970 then intermittently.
   Routine monitoring of dissolved oxygen, ammonia, BOD as a basis for river
   classification.
   1980. Two new classification schemes in addition to the above; one for
   rivers and canals, one for saline rivers and estuaries.
5. Harmonized Monitoring Scheme. Established 1974.
   To enable long-term trends in river water quality to be identified, and to
   meet international obligations.
6. Radioactivity.
   Two kinds of monitoring, collection and analysis of samples of waste
   (discharge monitoring), analysis of samples of environmental material
   (environmental monitoring).
7. Noise, Last national survey in 1972.
   1985–93, surveys by Open University Students.
   To provide trends in background noise levels.
8. Landscape Change 1984–86 (DOE and Countryside Commission).
   To monitor changes in the landscape in England and Wales.
9. Land Use Change. New Series, 1985.
   To provide an information source that is comprehensive and based on
   standard methods of definitions consistently applied across GB.

Source: HMSO (1986).

Agricultural and Food Research Council (AFRC) is an Environmental Change Network of long-term monitoring sites. The proposal is to designate a limited number of sites for which there are long runs of environmental data and either constant or well-recorded regimes. It is believed that such a Network would have considerable value, for example, in detecting the effects of global warming and the effect of land use change in Britain (see Chapter 14). Potential monitoring sites could include both nature reserves and some of the permanent vegetation plots, a register of which is held by the Institute of Terrestrial Ecology's Ecological Data Unit.

The Institute of Terrestrial Ecology

The Institute of Terrestrial Ecology (ITE) was established in 1973 and is one of about 15 institutes which make up the Natural Environment

Research Council. Monitoring programmes undertaken by ITE staff include the monitoring of heathland fragmentation (see p. 243) and monitoring pesticide residues in predatory birds. The Predatory Birds Monitoring Scheme (see Appendix II) is one of the longest running schemes of its kind in the world and results have been published in various scientific papers and a book by Cooke, Bell & Haas (1982).

The Ecological Data Unit (EDU) was established in 1984 within the ITE with the aim of collating data relating to ecological change and also to promote ecological monitoring. One of the first tasks of the EDU has been to prepare a register of permanent vegetation plots (Hill & Radford 1986). Surprisingly, the preparation of that register has shown that the range and number of projects (63) using permanent vegetation plots is greater than might have been expected (Table 9.2).

Biological Records Centre

Monitoring trends in populations, monitoring the distribution of a species and monitoring the extent of habitats and communities can be based only on good recording and survey methods. One example of a professional biological recording scheme which has, for many years, researched recording and survey methods for monitoring the distribution and abundance of various species is Britain's Biological Records Centre (BRC). Previously based at the University Botanic Gardens in Cambridge, the BRC staff and records were transferred in 1964 to the Monks Wood Experimental Station (Institute of Terrestrial Ecology) and today the BRC is incorporated in the NERC Environmental Information Centre (Sheail & Harding, in press; Perring, in press).

The BRC originated as the Distribution Maps Scheme of the Botanical Society of the British Isles (BSBI) which commenced in 1954 following the BSBI Conference in 1950 on The Study of the Distribution of British Plants. Some years later, after much research (particularly on the part of Franklyn Perring), came the publication of the first edition of *Atlas of British Flora* (Perring & Walters 1962). That atlas was not only the culmination of many years of data collection and research but was also a 'landmark' in the history of recording and survey methods. The *Atlas* is also a tribute to the skills of those who coordinated the many recorders throughout Britain.

One of the main aims of the originators of the BRC was the production of accurate distribution maps for the study of the history and biogeograhical relationships of the British Flora. That main aim continues today and distribution maps are available for several taxa including plants (Fig. 3.1). In Britain, the BRC based its recording on units of 10 km squares of the National Grid, partly because the squares are the same size irrespective of

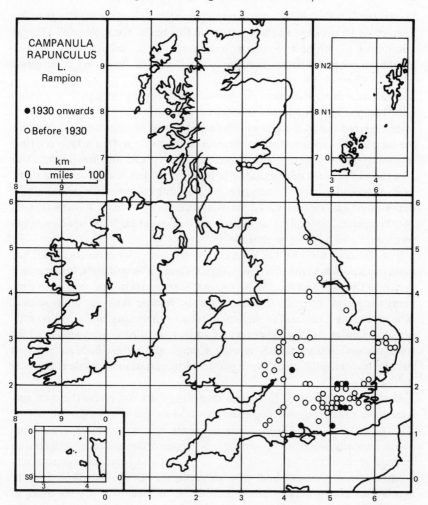

Fig. 3.1. An example of a species distribution map from the Atlas of British Flora. Redrawn from Perring & Walters (1962) with kind permission of the Botanical Society of the British Isles. Each circle or each dot represents one or more records of *Campanula rapunculus* in each 10 km square.

latitude and because they are marked on all ordnance survey maps. The status of a species expressed in terms of its occurrence in the number of 10 km squares can provide a basis for monitoring the change in status of a species. For example, the change in number of 10 km squares over a certain time period was one criterion on which the threat numbers (an index of conservation need) was based in Britain's first Red Data Book (Perring &

Farrell 1983). The species *Campanula rapunculus* in Fig. 3.1 has declined by as much as 10 per cent and has, partly on this basis, been allocated a threat number of 11 (highest recorded for any species in Britain is 13 and the maximum potential threat number is 15). Red Data Books are considered again in Chapter 16.

More recently, the BRC pioneered the mapping of national species distributions from a computer database and now the BRC has the largest computerized database for animals and plants in Britain. The database, in the long term, is the most valuable asset of the BRC, particularly in relation to biological monitoring, environmental assessment and modelling. The accuracy of species distributions is being recorded with more and more precision as advances in computerized mapping continue. To date the BRC data resource is impressive: 62 national recording schemes, information on 16 000 species, a total of around 5 million records, 5800 species maps published and 1200 in preparation.

The status of the BRC has been chequered but more recently (1989) has been incorporated within a newly established Environmental Information Centre at the Monks Wood Experimental Station, Huntingdon. This centre brings together considerable expertise in remote sensing, geographical information systems and ecological databases including biological recording. The Centre has the capability of addressing global problems, such as monitoring de-afforestation, climate change and desertification but also has the opportunity and experience to address national problems such as habitat loss.

Despite the extensive biological recording undertaken by the BRC and numerous other regional organizations and despite the extensive use of data from these recording schemes, Britain has no effective system for overall coordination of recording and monitoring wildlife. Interest in biological recording has never been so strong and regional organizations continue to be established. For example the Biological Recording in Scotland Campaign was founded in 1976 to meet the urgent need to promote and coordinate biological recording in Scotland. The need for national coordination of recording schemes led to the establishment of a working party on biological surveys in Britain. It was recommended in 1988 that a coordinating commission be established as soon as possible, under the lead of an appropriate national body (Berry 1988). This recommendation has been accepted at the time of writing and a commission is being established to oversee biological monitoring in Britain.

Nature Conservancy Council

The Nature Conservancy Council (NCC) was established in 1973 but was preceded by the Nature Conservancy, which had been established in 1949.

The statutory functions of the NCC, under the Nature Conservancy Council Act 1973, come under three main headings: establishment and management of nature reserves; provision of advice about nature conservation; and the support of research for nature conservation.

During the last 16 years, the NCC has been directly involved in, or has provided support for, a range of biological and ecological monitoring activities. For example, one current and particularly important national monitoring programme is the National Countryside Monitoring Scheme (jointly funded by the NCC and the Countryside Commission for Scotland). The details of this are given in Chapter 14. Other NCC monitoring programmes have been directed at particular taxonomic groups. For example, monitoring the abundance of butterflies (a joint project by the NCC and the Institute of Terrestrial Ecology) was started in 1976 after three years of preliminary trials (Pollard, Hall & Bibby 1986) and data is currently collected from 80 sites each year. The two aims with which the Butterfly Monitoring Scheme was set up have now been realized. It is now possible to show regional or national trends (Fig. 10.2) and it is possible to highlight the effects of management or lack of management.

In the first report of the NCC, it was made clear that the responsibilities of the NCC were not being fully achieved because of a lack of resources. Despite the importance of the work of the NCC, and despite the importance of the monitoring programmes (ten years' data on butterflies at some 80 sites represents invaluable information), the NCC has never received the support it deserves. Resources for the NCC have always been limited and at the time of writing there are plans to split the NCC into separate organizations for England, Scotland and Wales, linking the last two with the Countryside Commissions in those countries. These plans are a potential threat to the future success of nature conservation in Britain.

Rothamsted Experimental Station

Rothamsted Experimental Station, founded in 1843, is the oldest agricultural research station in the world and is funded largely by the Agriculture and Food Research Council (AFRC). Rothamsted is divided into several divisions including Agronomy and Crop Physiology, Biomathematics, Crop Protection, Molecular Sciences and Soils. A number of now classic long-term research programmes have evolved at Rothamsted, two examples of which are Lawes and Gilbert's Broadbalk Experiment (involving wheat) and Lawes and Gilbert's Park Grass Experiment (involving grass grown for hay). Both of these research programmes, which were initially ecological investigations into plant nutrition and elements limiting plant growth, have since led to a wide range of applied ecological and agricultural research.

In a recent major reorganization of agricultural and food research groups in Britain, 27 separate research stations and units funded by the AFRC in England and Wales were brought together into eight new institutions. Rothamsted Experimental Station together with Long Ashton Research Station near Bristol and other research groups now form the new Institute of Arable Crops Research.

Rothamsted Insect Survey

The distribution and abundance of insects has been monitored by agencies throughout the world for many years and there are many classic insect monitoring and census programmes (see Chapter 16). For example, commencing in 1950, there has been a winter moth census on five oak trees in Wytham Wood, Oxford (Varley, Gradwell & Hassell 1973). Monitoring the status of insect pest species has also been undertaken on a wide scale. As well as the obvious advantages in simply knowing where and what the pests are doing, monitoring insect pest species can provide information which may contribute towards better cost-effective control strategies.

The Rothamsted Insect Survey had its beginnings in 1959, first with moths and then later with aphids (Taylor 1986). From 1969 to 1984 this survey produced unique data; 364 daily records each year for a 1000 species from more than 100 sites; 10^8 observations to date. An insect pest monitoring service began in 1968 with the issue of weekly *Aphid Pest Bulletins*. These bulletins provided factual information on 32 aphid taxa flying over Britain as measured by 23 suction traps sampling at a height of 12.2 m (Taylor 1962). The bulletin was distributed each Friday and gave the number of aphids caught at the sampling sites, 5 to 12 days earlier. Today there is a network of suction traps and light traps throughout Europe and information on a variety of insect pests can be transmitted by electronic means rather than relying on the post. For example *Farmlink* (see Appendix ii), a farm management information service which is available via Prestel (British Telecom's Viewdata Network), can provide information from many advisory bodies including the Rothamsted Experimental Station.

The damage caused by some aphid species such as the Peach Potato Aphid (*Myzus persicae*), a species which transmits viruses and which can destroy potato and sugar beet crops, can be severe. For example in 1989, the Peach Potato Aphid accounted for at least £10 million worth of damage. On the other hand, chemicals used in agriculture, horticulture and forestry account for a large part of the budget and so need to be used only when necessary. Any information which can help to reduce this expenditure and indeed reduce the use of chemicals in the countryside should be welcomed.

Information which forecasts the location, type and abundance of the aphid species could, for example, contribute to a more effective use of chemicals. However, monitoring insect pests can be made very difficult because of the speed at which pests can travel and by the widely scattered distribution of the crops. Aerial sampling is an effective method providing data for distribution maps. A combination of ground surveys and forecasts from aerial surveillance can therefore provide an 'early warning' strategy and 'fine tuning' strategy so that prophylactic spraying is rejected in favour of target spraying. It is unfortunate that financial cutbacks have forced the insect pest monitoring service to use only 10 suction traps on a full-time basis and thus provide a less effective service.

The maps produced by the Rothamsted Insect Survey show the spatial and temporal distribution and the rate of change of distribution of both moth and aphid species. One example from Taylor (1986) shows the changing annual spatial distribution for moths (Fig. 3.2). The patterns revealed by these valuable data can not be explained only by population dynamics and require more research, for instance on the foraging behaviour.

Organizations monitoring birds

Throughout the world, many organizations monitor populations of birds and some of those organizations have international status. For example, the International Council for Bird Preservation (ICBP) is one example of an international organization which has played a major role in monitoring the status of many groups of birds, more recently the parrots. Some aspects of the work of national bird monitoring organizations are described in Chapter 11.

The British Trust for Ornithology (BTO) is the sole organization responsible for bird ringing in Britain and is the major contributor to bird census work via the Common Birds Census (which commenced with a pilot scheme in 1961) and related projects such as the Nest Record Scheme. However, there is a wide range of bird population surveillance in Britain (see Chapter 11). For example, the Game Conservancy administers the National Game Census (established in 1961), the aim of which is to monitor game bird population trends for species such as Red Grouse, Grey Partridge and Wood Pigeon. Details and an assessment of these schemes are in Chapter 11. Other bird monitoring programmes have been undertaken around Britain on Skomer, Fair Isle and other islands. The bird monitoring programmes on these islands owes much to the work of the Fair Isle Bird Observatory Trust (Williamson 1965, Fair Isle Bird Observatory Reports).

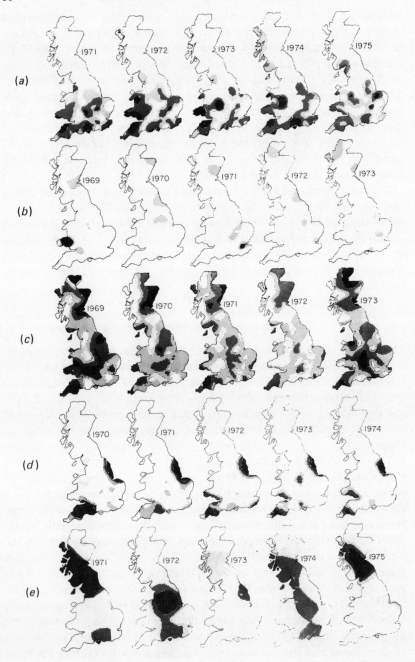

References

Berry, R. J. (ed.) (1988). *Biological Survey: Need & Network, A Report of a Working Party set up by the Linnean Society of London*. London, PNL Press.

Botkin, D. B. (1980). Life From a Planetary Perspective: Fundamantal Issues in Global Ecology. NASA Symposium on Global Ecology, Final report.

Brenneman, J. (1989). *Long-term Ecological Research in the United States. A Network of Research Sites 1989*. College of Forestry Resources, University of Washington.

Cooke, A. S., Bell, A. A. & Haas, M. B. (1982). *Predatory Birds, Pesticides and Pollution*. Merlewood, ITE.

Hill, M. O. & Radford, G. L. (1986). *Register of Permanent Vegetation Plots*. Abbots Ripton, ITE.

HMSO (1986). *DOE Digest of Environmental Protection and Water Statistics, No. 9*, HMSO.

Morris, H. L., Samiullah, Y. & Burton, M. S. A. (1988). Strategies for biological monitoring: the European experience. In *Metals in Coastal Environmens of Latin America*, ed. U. Seeliger, L. D. Lacerda & S. R. Patchineelam, pp. 286–92, Berlin, Springer Verlag.

Moss, D. (1990). The CORINE Biotopes Database. NERC *Computing*, March 1990, 24–5.

National Swedish Environmental Protection Board (1985). *Monitor 1985. The National Swedish Environmental Monitoring Programme PMK*. Stockholm, National Swedish Environmental Protection Board.

NOAA (1990). *Ocean Pollution Monitoring, Research, and Assessment*. A bibliography of NOAA-sponsored reports and publication. Ocean Assessments Division, Office of Oceanography and Marine Assessment National Ocean service, Maryland, NOAA.

Perring, F. H. (in press). BSBI *Distribution Maps Scheme – the First Forty Years*. NERC/BRC 25th Anniversary Vol.

Perring, F. H. & Farrell, L. (1983). *British Red Data Books, I, Vascular Plants*, 2nd ed. Nettleham, RSNC.

Perring, F. H. & Walters, S. M. (eds.) (1962). *Atlas of the British Flora*. London, BSBI/Nelson.

Pollard, E., Hall, M. L. & Bibby, T. J. (1986). *Monitoring the Abundance of Butterflies*. Research and Survey in Nature Conservation, No. 2. Peterborough, NCC.

Rhind, D. W., Wyatt, B. K., Briggs, D. J. & Wiggins, J. (1986). The creation of an environmental information system for the European Community. *Nachrichten aus Karten und Vermessungswesen*, Series 11, **44**, 147.

Schneider, G. (1989). The CORINE programme. *Naturopa*, **61** (1989), 15.

Sheail, J. & Harding, P. T. (in press). The Biological Records Centre – a pioneer in

Fig. 3.2. The spacial distribution of moth populations for five years of (a) *Spilosoma luteum*; (b) *Euxoa nigricans*; (c) *Xanthorrhoe flutuata*; (d) *Callimorpha jacobaeae*; and the aphid (e) *Aphis sambuci*. The isodensity contours have been plotted on an axis of grid coordinates using the log of the number of individuals caught in the network of light traps. From Taylor (1986) and reproduced with kind permission of Professor L. R. Taylor.

data gathering and retrieval. In *Biological Recording of Changes in British Wildlife*, ed. P. T. Harding & C. E. Appleby, London, HMSO.

Taylor, L. R. (1962). The absolute efficiency of insect suction traps. *Annals of Applied Biology*, **50**, 405–21.

Taylor, L. R. (1986). Synoptic dynamic, migration and the Rothamsted insect survey. *Journal of Animal Ecology*, **55**, 1–38.

Thomson, C. H. (1987). The Acid Drop Project: pollution monitoring by young people. *Biological Journal of the Linnean Society*, **32**, 127–35.

Varley, G. C., Gradwell, G. R. & Hassell, M. P. (1973). *Insect Population Ecology*. Oxford, Blackwell Scientific Publications.

Williamson, K. (1965). *Fair Isle and its Birds*. Edinburgh and London, Oliver & Boyd.

Part B

A BIOLOGICAL AND ECOLOGICAL BASIS FOR MONITORING

4

Elements of ecology and ecological methods

Introduction

THE AIM of this section is to describe some basic aspects of ecology which are relevant to ecological monitoring. It is not intended to provide a comprehensive introduction to ecology nor is it intended to consider any particular aspect of ecology in great detail. In this section we also consider some basic ecological sampling methods and analysis (further analytical methods are described in Chapters 5 and 6). Details of sampling equipment are not included here.

Data on the spatial and temporal distribution of organisms and data describing abundance of organisms have often been used as a basis for ecological monitoring. Studies of spatial distribution include a wide range of topics, such as the biogeographical distribution of a population of mammals or the distribution of insects on a single tree. Similarly, temporal distribution studies include many topics such as hibernation or aestivation of small mammals and circadian behaviour of some insects. Interactions between organisms, edaphic factors (factors relating to the soil), availability of resources and both physical and chemical factors are the important variables which affect distribution patterns. One approach to ecology is, therefore, the study of the factors which determine the distribution and abundance of plants and animals in space and in time. Such studies can provide a basis for biological and ecological monitoring.

Populations

One definition of a population is a group of organisms of the same species occupying a defined area. For example, we could be studying a population of Bluebells (*Endymion non-scriptus*) in a deciduous woodland or a population of Woodcock (*Scolopox rusticola*); see, for example, Fig. 11.7. Populations have various characteristics which are important in monitoring, the most basic of which are population size, dispersion (the distribution of individuals in space), population density, natality and mortality, and age-class or size-class distribution.

Population size

Over time, populations change in size, partly as a result of inherent population characteristics and partly in response to external factors. Although population size is widely used in monitoring, the population size of only a few species has been monitored over many generations or for long periods of time (see Chapter 16 for examples). This lack of detailed and continuous data from long-term population studies poses not insignificant problems in monitoring because, in the face of a multifactorial source of variability, it is difficult to distinguish trends due to pollution from those which are inherent, long-term cycles. The effects of harsh winters or dry summers on populations may not, for example, be easily researched without data from long-term studies (see Fig. 11.1).

Population distribution and limits of tolerance

All plants and animals have evolved to survive in different combinations of physical and chemical conditions. This can be demonstrated by looking at the distribution of animals in relation to chemical and physical factors, for instance the distribution of nereid worms in relation to salinity (Fig. 4.1). We can see that some species have narrow tolerances and some wide tolerance: *Nereis vexillosa* (one of the Ragworms) is confined to areas of high salinity whereas *Nereis diversicolor* has a distribution which extends from normal sea water to low levels of salinity. Organisms with narrow tolerances (or species which are sensitive to small changes in their environment) may be useful indicator species (see Chapter 5) because changes in distribution and abundance of these species may indicate environmental perturbations such as chemical pollution. Of course, all organisms are exposed to many kinds of abiotic factors and it is the combinations of factors which are important. For example, crops of Coconut Palms (*Cocos nucifera*) need a well distributed annual rainfall of about 50 cm for optimum production but these palms are limited in their geographical distribution by an intolerance to both low temperatures and arid conditions.

Knowing an organism's limits of tolerance to abiotic factors, it may then be possible to provide a basis for assessing effects of impacts on that

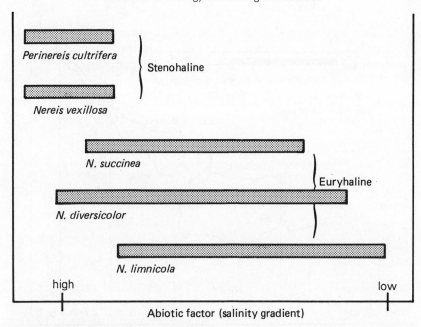

Fig. 4.1. The relative tolerances of Nereid worms to salinity. Redrawn from Vernberg & Vernberg (1970).

organism. Such assessments form the basis of the biological aspects of Environmental Impact Assessments (see Chapter 15). An example of impact assessment on individual species can be found in some of the preliminary work which has been undertaken as part of the assessment of the effects of a tidal barrage on the Severn Estuary (southwest England). The distribution and abundance of estuarine invertebrates in relation to salinity levels has been assessed by the Nature Conservancy Council (Mitchell & Probert 1981). Their reports predict that the barrage will alter salinity levels in the estuary and consequently the distribution (and abundance) of the estuarine invertebrates will change (Fig. 4.2). The estuarine invertebrates are important food sources for shore birds and so changes in salinity patterns brought about by the construction of the barrage will also affect the distribution and abundance of the birds.

Distribution patterns

Distribution patterns are many and varied and some interesting geographical distribution patterns have been found to occur in relation to latitude. That is, for some taxonomic groups such as lizards, the number of species

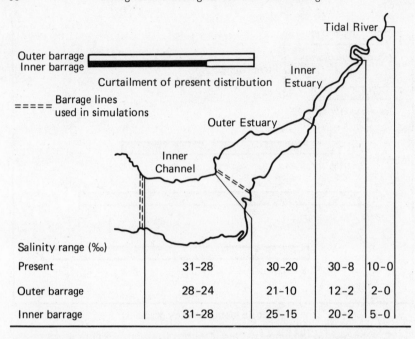

Common benthic species:

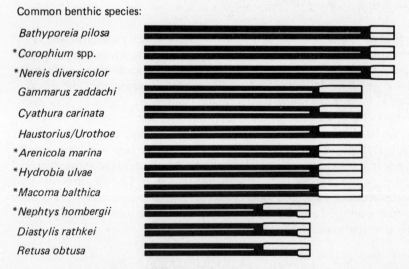

Fig. 4.2. Predicted effects of salinity changes on the distribution of benthic species in an estuary. An asterisk (*) indicates important prey species for birds. Redrawn from Mitchell & Probert (1981) after Warwick & Uncles (1980).

generally declines with latitude so that more species are found closer to the equator and fewer species at high latitudes. Another interesting pattern has sometimes been found when comparing species richness on peninsulas with mainland fauna. Simpson (1964) noted that some mammal taxa on peninsulas were fewer in number than those inhabiting equal-sized mainland regions. On peninsulas themselves, there is sometimes a decrease in species richness from base to tip. For example, Means & Simberloff (1987) found that for Florida's amphibians and reptiles, 48 of 108 species inhabiting the peninsula failed to reach the peninsula tip. Another interesting aspect of spatial distribution is isolation; some species are isolated in caves, some in alpine regions and some in lakes. Patchy distribution of resources may result in a degree of isolation such as some species of insects on inflorescences or on trees and marine organisms on intertidal boulders. This patchy distribution may present some difficulties when attempting to estimate population density and it is essential that the sampling unit is able to detect such patchy distributions.

In theory, the spatial distribution patterns of a population can be uniform (regular), random or aggregated (Fig. 4.3). Patterns of dispersion can be quantified by taking samples and calculating the mean and variance (s^2) (that is, the square of the standard deviation). If the individuals in a population are dispered uniformly then the variance (s^2) is zero because the number of individuals in a sampling unit would be equal to the mean value (average number of individuals for the sampling replicates). If the population had a random distribution then the mean and variance would be equal. If aggregation occurs, then the variance becomes much greater

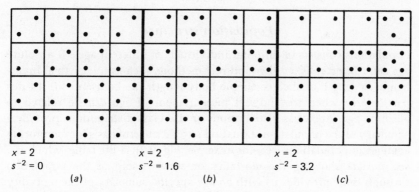

$x = 2$ $x = 2$ $x = 2$
$s^{-2} = 0$ $s^{-2} = 1.6$ $s^{-2} = 3.2$
(a)   (b)   (c)

Fig. 4.3. Spatial distribution of individuals in hypothetical populations. (a) Regular pattern with no variance; (b) approaching a random distribution where the variance would equal the mean if there was a perfectly random distribution; (c) aggregation of individuals so that the variance is much greater than the mean. From Spellerberg & Pritchard (1984).

than the mean as the individuals become more and more aggregated. The spatial dispersion of a population is seldom random or regular but is frequently aggregated. This is not surprising, given the uneven distribution and concentration of environmental factors.

Population density

Population density is measured as the number of individuals per unit area or unit volume, but methods of estimation of population density are very varied according to the taxonomic groups being monitored. Measurements of absolute density depend on counts of individuals in areas which can be measured. For example, population density of birds can be calculated on the basis of numbers of individuals in an area of known size. For relatively immobile species, the use of quadrats (squares of a known size) provide a basis for measuring density in relatively small areas, the results of which can be extrapolated to larger areas. For example, quadrats of a metre square could be used to calculate the population density of grasses and herbs or a quadrat of $0.25 \, m^2$ could be used to determine population density of barnacles on a rock. Larger quadrats are used for surveys of woodlands (see, for example, Kirby 1988). Measurements of relative population density can be undertaken using a variable that has some relatively constant relationship with total population size. Relative density of mammals can, for example, be based on faecal counts if the rate of defecation and the number of faecal pellets in a certain area is known. Plant percentage cover in a known area (or in a quadrat) has been widely used as an estimate of relative abundance in monitoring studies.

Population structure

Populations consists of different individuals of different ages. If absolute age can't be established then indirect methods such as tree girth at breast height or length of shell or simple body length can be useful substitutes particularly when the indirect measurement of age can be related to absolute age. Age-class distributions or size-class distributions provide a summary of the population status and so give a useful basis for monitoring changes in populations (see, for example, Fig. 9.2). A life table, which is a set of data showing the mortality or survivorship of the population through time, provides us with an age-specific summary of the mortality rates which occur in the population. Life tables are constructed from data on age intervals, number of survivors at each age interval and number dying at each interval. The rate of mortality and mean expectation of life can then be calculated from the life table.

Communities and ecosystems

An ecological community is an assemblage of interacting populations of different species in a particular area or habitat. An ecosystem consists of the ecological community and the abiotic environment with which it interacts in a dynamic and complex way. Some communities such as a permanent pasture are man-made and the species composition is dominated by one or two species. Other communities such as a fragment of a mixed deciduous woodland may not be man-made but there may well be examples of previous management of some tree species by way of coppicing. An ecosystem such as an agricultural ecosystem or woodland ecosystem includes the biological community, the soil, the air and the water all interacting and functioning together as a whole. One rewarding way of studying ecosystems is to trace the nutrient and energy flow through the system.

Energy flow and nutrient cycling

Temperature, light, water and nutrients are important variables in communities. Plant species composition, for example, is particularly dependent on soil structure and levels of various soil nutrients such as nitrogen, potassium and phosphorus. The importance of nutrient levels can be seen in the form of plant community indicators, that is the species composition reflects the soil conditions (see Chapter 5).

Within ecosystems there is a continuous turnover of mineral nutrients and a continuous flow of energy. The plants and animals in a community, together with the abiotic environment, through which nutrients are recycled, constitutes an ecosystem. A diagrammatic representation of potassium circulation is shown in Fig. 4.4 and it is interesting to see that there are noticeable differences in the relative amounts of potassium circulated in pine woodlands and oak woodlands.

Productivity

The change in total dry weight of organisms (biomass) in an ecosystem over time is called productivity. Energy fixed per unit time by a community through the process of photosynthesis is called gross primary production. Net primary production is the gross primary production minus the energy used in plant respiration, and is thus a measure of the energy available for the consumers. Net community production (NCP) is the net primary production minus the energy used by the consumers and represents the absolute increase in community energy (or biomass) in a given period of time. A climax community commonly has an NCP of zero because primary production equals total consumption.

Relationships between different trophic levels are rarely simple. Harvesting, exploitation or management of one group of organisms can directly or indirectly affect other groups of organisms at other trophic levels. The use of pesticides, for example, may have complex implications

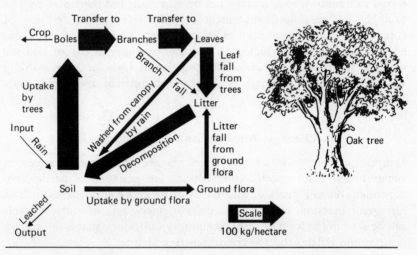

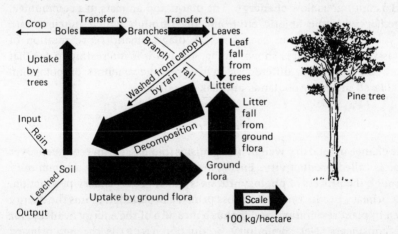

Fig. 4.4. Diagrammatic representation of potassium circulation in two adjacent woodlands, oak woodlands and pine woodlands. The thickness of the arrows indicates the magnitude of the flow of potassium. Redrawn from Ovington (1965).

for an ecosystem, as exemplified by the accumulation of chemicals at high trophic levels. The thinning of egg shells of birds of prey, brought about by an accumulation of DDT, has already been mentioned on p. 11.

Trophic levels and food webs

In terms of ecosystem functioning, there are three basic groups of organisms in an ecosystem: the primary producers, the consumers and the decomposers (Fig. 4.5). Energy from the sun is utilized at the first trophic level by the green plants (producers or autotrophs) for the basic metabolic process of photosynthesis, producing carbohydrates and oxygen. Although energy from the sun is the primary source of energy for most ecosystems there is an interesting exception. Some organisms derive energy from energy-rich inorganic molecules such as those found around

FLOW OF ENERGY

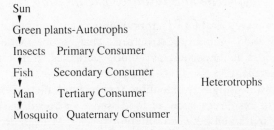

Fig. 4.5. Schematic representation of energy flow in a hypothetical food chain. At each transfer of energy, as much as 90 per cent of the stored energy may not be available to the next stage. In most ecosystems the biomass of the producers (autotrophs) is greater than that of the primary consumers, and the biomass of the primary consumers is greater than that of the secondary consumers and so on.

volcanic vents on the sea floor. The heterotrophs at other trophic levels include the herbivores or primary consumers, the secondary consumers (carnivores) and the tertiary consumers. The complex life cycles of some organisms mean that a species may not always occupy the same trophic level; the hypothetical Mosquito in Fig. 4.5 is a quaternary consumer as an adult but a secondary consumer during the larval stage. Similarly, many amphibian species occupy different trophic levels at different stages of their life cycle.

Community structure

The structure of a biological community can be described in various ways, based on variables such as biomass at different trophic levels, the species composition, life forms and relative abundance of species etc. All of these variables and many other measures of community structure are widely used in ecological monitoring. For example, one fundamental aspect of plant and animal community structure and one which has played an important role in monitoring changes in communities is the association between two or more species. A description of how species associations can be quantified is described below and more details of analysis can be found in texts such as Digby & Kempton (1987). The abundance of certain indicator species (Chapter 5) and also a variety of biotic indices (see Chapter 8) have been used as a basis for describing a community. Quantitative or qualitative descriptions of biological communities over time have usefully been used as a basis for monitoring change in those communities.

Niche and resource partitioning

Biotic factors as well as abiotic factors are important in determining species composition and community structure. That is, interactions between species are just as important as abiotic variables, if not more so, in community ecology.

One very interesting way of looking at species interactions is to examine how different resources such as food, space or roosting sites etc. are divided or partitioned between different species. We can do this by first looking at the niche of a species. One way of thinking of a niche is the sum of all the relationships between the species and its environment; alternatively we can think of the niche as the activity range of each species along every dimension of the abiotic and biotic environment (Hutchinson 1959). In isolation, plants and animals are able to exist in a wide range of environmental conditions and that range of conditions defines what is known as the fundamental niche. Placed outside the fundamental niche, the plant or animal will not survive. In natural communities a species has a realized niche, that is in the presence of competition and predators, not all resources may be used to that extent when the species was isolated and in those circumstances we refer to the realized niche of a species. In captivity, for example, a lizard might feed on several groups of invertebrates in roughly equal proportions to the abundance of those invertebrate groups. In natural communities and living in competition with another species of lizard with similar food preferences, that lizard species may take a large proportion of one group of invertebrates and few of the other groups. The change in the one dimension of the lizard's niche is simply part of the realized niche in the face of competition.

Each dimension in the abiotic and biotic environment can be thought of as a dimension in space. If there were n dimensions, the niche is described in an n-dimensional space, of which the species occupies a certain defined volume. The concept of niche width (or breadth) and niche overlap (Fig. 4.6) can usefully be used to quantify resource partitioning.

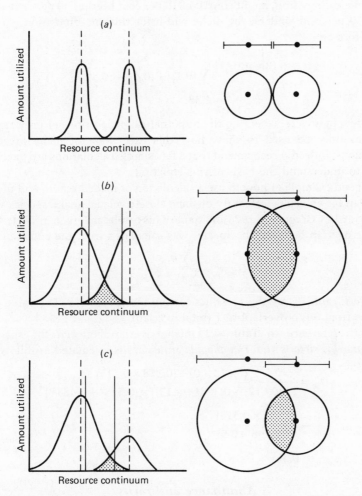

Fig. 4.6. Diagrammatic representation of niche overlap: (a) contiguous but no overlap; (b) overlapping of equal breadth; (c) overlapping of unequal breadth. Modified from Pianka (1969).

For example, niche breadth can be quantified using a diversity index (see Chapter 6) such as Simpson's:

$$B \text{ (Breadth)} = \frac{1}{\sum\limits_{i=1}^{s} P_i^2}$$

where s is the number of different resources, P_i is the frequency of utilization of the ith category. For example, a lizard's prey (the prey being one niche dimension) might consist of 10 per cent beetles, 40 per cent ants and 50 per cent spiders: the niche width for this one dimension would therefore be:

$$B \text{ (Breadth)} = \frac{1}{\sum (0.1)^2 + (0.4)^2 + (0.5)^2}$$

$$= 2.38.$$

A few figures representing the hypothetical proportions of prey types can usefully be used to show how this expression of niche breadth functions. Indeed, I recommend that a few simple calculations are the best way to understand the basis of this equation.

A good example of how to use a calculation of niche breadth and niche overlap for an ecological study of some closely related snake species was described by Gregory in 1978 and some of his results are given in Table 4.1. Niche overlap between the species was calculated from the equation:

$$O_{jk} = \frac{\sum P_{ij} P_{ik}}{\sqrt{(\sum P_{ij}^2 \sum P_{ik}^2)}}$$

where P_{ij} and P_{ik} are the P_i values for the species j and k. The value O_{jk} varies from 0 (no overlap) to 1 (total overlap).

With reference to Table 4.1, niche overlap between the species *Thamnophis sirtalis* and *Thamnophis ordinoides* is calculated as follows:

$$O_{jk} = \frac{\sum (0.612 \times 0) + (0.358 \times 0.515) \ldots}{\sqrt{[(0.612^2 + 0.358^2 + 0.15^2) \times (0.515^2 + 0.485^2)]}}$$

$$= \frac{(0.358 \times 0.515)}{\sqrt{(0.498 \times 0.5)}}$$

$$= 0.36$$

Abundance and rarity

In some biological communities, there may be some species which are rare because of our exploitation of wildlife or because our activities have had an

Table 4.1. *Niche breadth and niche overlap in three species of the snake* Thamnophis

Food type	T. sirtalis f_i^*	T. sirtalis p_i	T. ordinoides f_i	T. ordinoides p_i	T. elegans f_i	T. elegans p_i
Amphibians	41	0.612			2	0.015
Earthworms	24	0.358	67	0.515	5	0.038
Slugs			63	0.485	60	0.458
Fish					41	0.313
Mammals					17	0.130
Birds	1	0.015			2	0.015
Leeches	1	0.015				
Reptiles					4	0.031
Σ	67	1.000	130	1.000	131	1.000
B		1.987		1.998		3.054
O_{jk}		0.367				
				0.057		
						0.597

Note: mean $O_{jk} = 0.340$.
*f, absolute frequency.
From Gregory (1978).

indirect, harmful effect on the wildlife. However, in nature, some species are naturally rare and some are common. It is more realistic to say that most species are rare (represented by few individuals) and few are common (represented by large numbers of individuals). The commonness and rareness of species is an important and interesting phenomenon which is basic to our understanding of the use of species richness, species diversity and other variables used in monitoring (see Chapters 6, 7 and 8). There have been several attempts to describe the phenomenon in mathematical terms: for example, Fisher, Corbet & Williams (1943) found that the data were best fitted by a log series which implies that the greater number of species have a minimal abundance. The relationship between the number of species in the sample and the number of individuals in the sample gives a diversity index, independent of the sample size and has been used in monitoring changes in communities (see Fig. 6.2).

The description provided by Fisher *et al.* (1943) has been found not to apply to all communities and Preston (1948) introduced the idea of expressing the number of individuals in a sample on a geometric scale rather than an arithmetic scale. In many cases this results in a normal (bell-shaped) distribution known as log normal. In 1962, Preston noted that the log normal distribution had a characteristic appearance or configuration which he termed the canonical distribution. The word 'canonical' was chosen for ecological purposes as a rough analogy to a state of statistical equilibrium previously observed in mathematical physics; 'log normal distribution of the abundances of various species (or genera, families, etc.) whose individual curve terminates at its crest'. That there is no apparent theoretical basis for a log normal distribution of relative abundance of species is a major criticism. An alternative approach has been via the use of information theory which asks the question how easy is it to predict correctly the species of the next individual collected? This can be measured by a Shannon–Weiner function which is also used as a measure of species diversity (explained on p. 123).

Ecological succession

The development of a community over time, or the process of change, is called ecological succession. Although succession is a process of continuous change it is possible to recognize a series of phases (seral stages) through which the community develops. Succession can be viewed as the development of a community from its inception (the pioneer stage) which is then replaced by a whole series of communities. The whole sequence is called a sere and each successional stage can be considered to be a community in its own right. The final or mature community is called the climax community. In general there is a low species richness in the early stages of succession and species richness increases as the climax community is approached.

   Establishment and development of communities in newly formed habitats, for example on newly formed sand dunes, is termed primary succession. Where natural communities have been destroyed and there is recolonization, then this is called secondary succession. In general the process of ecological succession includes soil development, increasing structural complexity, and an increase in biomass. In addition there are changes in species composition or assemblages until the climax is reached. In terrestrial habitats the climax stage is usually dominated by long-lived plants. In some circumstances, the process of natural change may be deflected or arrested by man's activities such as by burning or cutting. For example, the heathlands of Europe are managed communities and natural change to woodland has been arrested resulting in a plagioclimax community.

Ecological methods

Qualitative and quantitative surveys

What species are there? Where are they? How many are there? These are the basic questions in field surveys. Finding out what species are present in order to build up a picture of species composition requires less rigorous sampling than does estimating species abundance. But in order to establish the species composition of an area we need to know how many samples should be taken, or how many quadrats, or how many sweeps of the net, or how many transects.

To find out how many species are present, a useful approach is to keep a cumulative record of the number of species recorded (along transects or in quadrats). Eventually the chances of finding a new species in the sampling area will diminish to a low probability and that low probability indicates that thorough sampling efforts have been established (Fig. 4.7).

Random sampling is a commonly accepted method but because sampling may be carried out over an area with more than one type of community, stratified random sampling is used. Stratified random sampling simply means that sampling is divided on the basis of a subjective classification of the area. For example, stratified sampling of an area of plantations and heathlands would result in different sized quadrats and a different number of quadrats to ensure satisfactory sampling in each community (Kirby 1988).

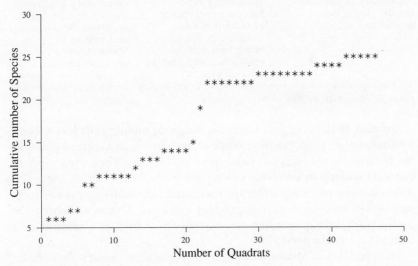

Fig. 4.7. Hypothetical species–area curve obtained by cumulative sampling.

Table 4.2. *A quadrat and data collection model for monitoring endangered plants at the Indiana Dunes National lake shore*

Objectives
(1) Design quadrats compatible with plant life form, population size and population distribution.
(2) Collect data on population characteristics that will allow monitoring of specific population changes.

Population size and distribution	Mapping and Quadrat design	Data Collection
Locally rare small woody plants, annual or perennial herbs, or cespitose grasses and sedges.	Map genet distribution by sq. dm, use multiple meter sq. quadrats over entire population.	Ramets/genet, no. flowers or fruit/ramet or genet. Each genet = a data point.
Locally distributed large woody plants or coarse herbs, sedges or grasses.	Map ramets or genets by smallest possible meter sq. unit. Map distance and direction from circular plot center.	Ramets/genet, no. flowers or fruit/ramet or genet. Each ramet or genet = a data point.
Small dense populations of woody plants, sedges or grasses in which mapping and quantification of individuals is not facilitated.	Map cover by smallest usable unit area, limit grid size to 100 sq. m; or map distance and direction from up to 1/10 ha plot centers.	Ramet or genet frequency, density, cover-class, or size-class/unit area; no. flowers or fruits/ramet or genet/unit area.
Extensive population of community dominant woody plants, herbs, grasses and sedges.	Map extent of community type. Sample by stratified random meter sq. quadrats or intercept along transects; or establish study plots at selected points.	Determine relation to community structure. Sample density, cover or size-class/unit area, and flowers or fruits/ramet or genet/unit area.

From Bowles *et al.* (1986).

A useful, basic set of parameters for designing sample plots was devised by Bowles *et al.* (1986) in their work on monitoring endangered plants at the Indiana Dunes National Lake shore (Table 4.2). They used plots or quadrats ranging in size from one square metre to 100 square metres to collect data on population characteristics and map such units as ramets (a unit which may have an independent existence if separated from the parent plant) and genets (the genetic individual such as a seedling or a clone) (Kays & Harper 1974).

The precision of absolute estimates of population density depends on several factors, including the temporal and spatial patterns of dispersion of the organisms, the methods used for sampling and of course observer bias (a subject discussed more fully in Chapter 11). A decision about the sampling effort required for a monitoring programme can not easily be made without at least some information on the spatial and temporal distribution of the organisms and their behaviour, including aspects of phenology. The sample size should provide adequate and statistically reliable data but at the same time we must be aware to the possibility that an accumulation of large amounts of material as a result of large sampling efforts may incur high costs and require long periods of time to undertake the sorting, identification and analysis. Basically, the sample replicates should be large enough to include at least one individual of each species (in the area being examined).

Each sample replicate, whether it be via a quadrat or some kind of sophisticated sampling apparatus such as a suction trap (see Rothamsted Insect Survey on p. 56). will give a different result in terms of number of individuals of each species. Therefore, a number of sample replicates are required and a statistical analysis of the replicates will give an indication of precision. On a very simple basis, one equation for calculating the number of sampling units required for a given error of the mean is as follows:

$$n = (ts/D\bar{x})^2$$

where t is the Student's-t for the required probability, s the standard deviation and D is a given relative error of the mean (x) in a random sample. For example, if a preliminary survey of herbs in a meadow gave a mean of 12 individuals of one species per sample replicate and a standard deviation of 3, then for an error of 10 per cent with a probability of 95 per cent (t roughly equals 2) the number of sample replicates would be:

$$(2 \times 3/0.1 \times 12)^2 = 25$$

There is really no wholly satisfactory way of determining the number of sample replicates required (Southwood 1966, Lewis & Taylor 1967, Chalmers & Parker 1986) but according to Taylor (1961) the variance (s^2) of a sample varies with the mean (m) according to the following power law:

$$s^2 = am^b$$

where a is largely a sampling factor and b is a constant for a given species in certain conditions. The relationship can usefully be expressed logarithmically as follows:

$$\log s^2 = \log a + b \log m$$

where $\log a$ and b are the intercept and regression coefficient.

The sample size (N) for a given level of precision (p) can then be calculated as follows (Vickerman 1985):

$$\log N = (\log a - 2\log p) - (2 - b)\log m$$

Some publications give estimates of sample size required for various taxonomic groups (Hales 1962, Vickerman 1985). For example, in a study of environmental monitoring using bird and small mammal populations in four Utah cold desert habitats, Steele, Bayn & Grant (1984) investigated the sampling necessary to make an adequate estimate of characteristics of bird communities. They found that estimates of bird abundance, species richness and species diversity (Shannon–Wiener Index; see p. 123) could be obtained from three repetitions of 2 km of transect. Small mammal abundance on 12×12 trapping grids also required three repetitions to estimate abundance but species richness and species diversity required at least four repetitions.

Frequency of sampling and location of sampling

Monitoring requires that information is collected at intervals throughout time and the frequency of data collection is crucial to the success of the monitoring programme. Different species have different cycles of activity (including seasonal, diurnal and tidal) and obviously infrequent data collection may lead to misleading interpretations whereas an excessive time based data collection (particularly destructive sampling) may be harmful to the ecosystem and also may not be cost-effective. A knowledge of the organism's life cycle and conservation needs will also determine the frequently of data collection (see, for example, Table 16.1).

In essence, the frequency of sampling depends partly on the objectives of the monitoring but the deciding factor should be based on a requirement for statistically useful data. Here we need to look for some guidance from time series analysis and the statistical analysis of periodicity (see, for example, Bloomfield 1976). For example, the sampling frequency in some monitoring programmes is at least twice the highest frequency of the event being monitored and thus when a large part of the variance in a variable is found in the high frequencies, that variable would not be suitable for monitoring. Govaere et al. (1980) adopted this approach based on the Nyquist frequency in their ecological monitoring of benthic communities in the North Sea. The Nyquist frequency is one-half the sampling rate; that is, there are two samples per cycle of the Nyquist frequency, the highest frequency that can be observed. Bloomfield (1976) and other texts on time-series analysis should be consulted for more detailed explanations.

The location of sampling replicates is usually 'random' and a simple way to achieve this is to establish a grid, with random numbers on the coordinates, then sample at points based on a random selection of coordinates. An alternative to a grid system is a transect and a good example of its use has been in connection with monitoring fauna and rocky shores, especially in relation to oil spill incidents. The ecological effects of oil spills have attracted much attention in recent years; for example the Field Studies Council Research Centre, previously the Oil Pollution Research Unit (see pp. 37 and 183) has established oil spill monitoring programmes based on permanent transects. Once established, subjective assessments (abundant, common, frequent, occasional, rare) are made of the abundance of organisms at points on the transect (Hiscock 1985, Little & Hiscock 1989). The Field Studies Council Research Centre has refined the use of an abundance-scale technique on uninterrupted belt transects in a number of ways; for example, when monitoring barnacle populations, the estimation of abundance scores has been improved with the use of counts from small quadrats (see Table 15.1). More detailed information about ecological methods for sampling can be found in large texts such as Southwood (1966) or smaller texts such as Chalmers & Parker (1986) or Gilbertson, Kent & Pyatt (1985).

Population size

The population size of many species of animals and plants is difficult to measure directly, let alone estimate. In general, the higher plants are easier to count than some vertebrate animals and most difficult are the small, microscopic animals for which specialized counting techniques have been devised (see, for example, 'Detectors and exploiters' in Chapter 5). The behaviour of some animals such as the Emperor Penguin (*Aptenodytes forsteri*) has on some occasions made counting relatively easy (Fig. 4.8). At the other extreme, there are some conspicuous species, such as Guillemots, which are not easily counted in a breeding colony because of the high density of birds in the colony (see Fig. 4.9, and p. 199 in Chapter 11).

Observer accuracy when counting is a subject which has been well analysed and it has been found that even seemingly easy tasks such as aerial counts of large mammals are subject to observer's inaccuracy. Fig. 4.10, from one of Caughleys's papers, can be used to simulate aerial counts and demonstrate errors arising from observer accuracy. Caughley (1977) addresses this and other related counting problems in his book.

In some cases, such as many common bird species, the population is so large or so widely dispersed that it is impractical to count all individuals. In these circumstances it may be possible to establish a population index, that is a regular census of small sample populations. Changes in the index may reflect changes in the total population and indeed this method has been used successfully by the British Trust for Ornithology in their Common Birds Census (described on p. 206 *et seq.*).

The well established capture–recapture method of estimating population size has been found to be successful for a few taxonomic groups. For example, it has been possible to make fairly accurate estimates of the size of some small mammal populations by first capturing a sample of individuals, marking them or recording individual, characteristic markings, then releasing them. In a subsequent sample, the ratio of previously caught individuals to the total sample size is hypothetically the same as the ratio between total individuals marked and the total population size. There are several models for this method (Southwood 1966), one of which is based on the ratios between marked and unmarked animals:

$$\frac{\text{number of marked animals in sample}}{\text{total caught in sample}} = \frac{\text{number marked in total population}}{\text{total population size}}$$

The total population size (P) can be calculated:

$$P = an/r$$

where a is the total number of animals marked, n is the total number of

Fig. 4.8. Emperor Penguins (*Aptenodytes forsteri*) walking single file at Ross Island, Antarctica.

Fig. 4.9. Part of an enormous colony of Brunnich's Guillemots (*Uria lomvia*) on Coburg Island, Canada. (Guillemot photograph by T. R. Birkhead.)

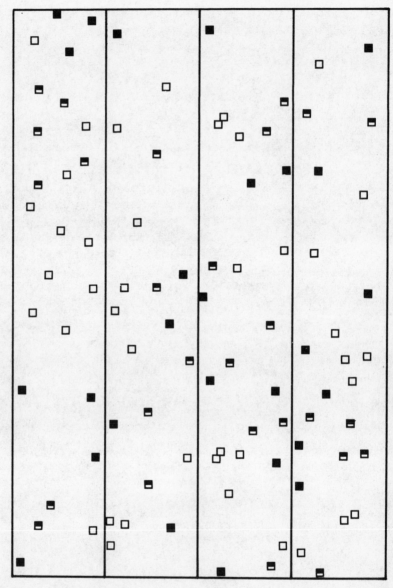

Fig. 4.10. A hypothetical distribution of organisms within a sampling grid. Simulation of error in an aerial census can be undertaken by counting the individuals

individuals caught in a subsequent sample and *r* is the total number of marked animals recaptured.

The various models of capture–recapture are based on a number of assumptions including the following: the behaviour of the captured and marked animals does not change as a result of handling and marking, the marked animals become completely mixed in the whole population, the population is sampled in a random fashion.

Species associations

Species composition, relative abundance, dominance and species associations are all important community characteristics which have been applied to monitoring changes in communities. Changes in species associations have been especially useful in monitoring and there has been much research on ways of measuring and describing species associations. In surveys of grass swards, the quadrat size needs to be given careful consideration otherwise associations may not be detected. Kershaw (1964) gives a nice hypothetical illustration of this (Fig. 4.11). An association exists between species A and B, whereas negative associations exist between A and C and between B and C. Sampling with different quadrats will give different results, quadrat 1 will show no associations, quadrat 2

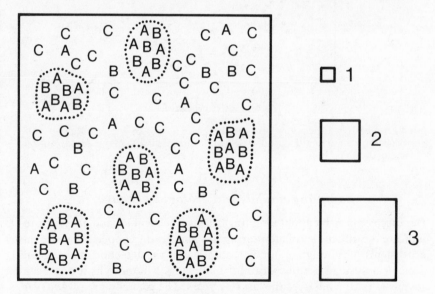

Fig. 4.11. Relationship between quadrat size and trend of association among three hypothetical species, A, B and C. Redrawn from Kershaw (1964) with permission of Hodder & Stoughton.

will show the expected associations between A and B and quadrat 3 will contain all three species.

One way of overcoming this problem of quadrat size is to use a form of nearest-neighbour analysis to assess an association between two species. A random sample of each species (A and B) is taken in turn and on each occasion the nearest species (A or B) is recorded. These data are then set out in a contingency table showing the number of occasions species A was nearest to species A or B and vice versa. A simple Chi Square provides appropriate statistical analysis. Negative associations can be shown very nicely by way of percentage cover of two species in a series of quadrats (Fig. 4.12). Further and relevant analysis, by way of ordination and correlation statistics is described on p. 136.

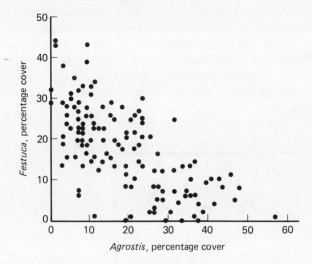

Fig. 4.12. Negative associations between the grasses *Agrostis* and *Festuca* in grasslands in the north of England. Redrawn from Kershaw (1964) with permission of Hodder & Stoughton.

Sampling models and ecological change

Our discussion so far about sampling has been elementary and does little to indicate the difficulties of monitoring the complexities of perturbations on ecosystems. The design of optimal sampling for the detection of change in ecological systems requires careful modelling, a topic which has been reviewed fairly well over the last few years. The common use of ANOVA (analysis of variance) in the analysis of data from monitoring programmes (see, for example, Steele *et al*. 1984) can be used to illustrate the complexities involved. Millard & Lettenmaier (1986) give a good example of monitoring based on ANOVA prior to their account of the development of an optimal design of biological sampling programmes.

The moderately complex example used (Millard & Lettenmaier 1986) is based on an hypothetical electricity generating station impact assessment. The aim would be to monitor changes in population density of selected species during the operation of the power plant and if observations were taken at various sites in the area before and after the development, they suggested the following equation for the basis of a model:

$$Y_{jklmn} = U + A_j + B_k + C_l + AB_{jk} + AC_{jl} + BC_{kl} + ABC_{jkl} + N_{jklm} + E_{jklmn}$$

where

Y_{jklmn}	= the observed density for the nth replicate on the mth sampling occasion within season l at station j for perturbation k,
U	= the overall mean density, over stations, perturbations, and seasons,
A_j	= the mean effect of station j,
B_k	= main effect of perturbation at level k,
C_l	= main effect of season l,
AB_{jk}	= station by perturbation interaction effect at level jk,
AC_{jl}	= station by season interaction effect at level jl,
BC_{jl}	= event status by season interaction effect at level kl,
ABC_{jkl}	= station by event status by season interaction effect at level jkl,
N_{jklm}	= effect due to sampling occasion m within season l for station j and perturbation k,
E_{jklmn}	= random error in the nth replicate on the mth sampling occasion within season l at station j for perturbation k,
j	= generating station,
k	= perturbation (1, before, and 2, afterwards).
l	= season,
m	= sampling occasion,
n	= replicate.

Models based on this equation for detecting ecological change have met with criticism especially with regard to field sampling designs and whether or not true replicate samples can be obtained. In addition, Millard & Lettenmaier (1986) felt that ecological monitoring should be designed to assess the effect of a particular perturbation upon a particular ecosystem, a view which contrasts with the classical scientific design in which the objective would be to assess the average effect on a number of ecosystems.

The optimal design model suggested by Millard & Lettenmaier (1986) incorporates not only assessment of effects of perturbations on an ecosystem but also the costs incurred in detecting those effects. Optimization takes two forms, minimizing of sampling costs for a given detectability of change and maximization of power for a fixed cost. This illustrates the complexity of monitoring ecosystems and also reminds us that the application of monitoring may extend beyond ecological parameters to economic constraints.

Data collection and recording

Methods

Although many data recording methods can now be undertaken with very sophisticated equipment, there is still much field ecology which can be undertaken in the most rigorous fashion with the most basic of equipment. A fundamental rule to remember is that all sampling and recording has its limitations and that sampling equipment is always biassed towards certain taxonomic groups. Even the most basic of all invertebrate traps, the pitfall trap, does not catch a representative sample of the fauna because some groups are more likely to be caught than others (Southwood 1966, Halsall & Wratten 1988). This is not to say that all recording and sampling equipment is unsatisfactory; the limitation and bias need only be recognized and if possible quantified.

While simple sampling techniques can be undertaken by hand with the most simple of equipment, continuous sampling devices such as the suction traps used for aerial invertebrates by the Rothamsted Insect Survey (p. 56) have obvious advantages, particularly with regard to analysis of temporal patterns of distribution (see Fig. 3.2). Automation of sampling in conjunction with continuous recording of environmental variables by way of data logging equipment has provided new opportunities for monitoring complex systems.

There are some biological and ecological monitoring programmes which rely on repeated data collection from exactly the same site (see for instance the example described on p. 284). These sites can be marked in many ways but despite ingenious methods of marking, the methods for long-term marking of sites (including so-called permanent quadrats) could usefully be explored much further. An obvious method for marking woodland quadrats is with the use of posts (see, for example, Fig. 9.3) but these posts deteriorate after a number of years and are prone to vandalism. One alternative method is to use insulated metal markers in the ground (or insulated metal strips marking the whole quadrat) which are then located by metal detectors. The combination of metal markers and posts (perhaps supplemented with photographs) is the least which can be done to ensure that samples are taken from exactly the same quadrat. Metal quadrats can also be designed to identify individual plants in actuarial monitoring (Bradshaw 1981).

There is a large variety of equipment available for ecological sampling and recording. This equipment ranges from standard colonization units for monitoring water quality in rivers to light traps used for catching moths. It would not be practical even to attempt to describe this equipment here but the following excellent texts are worth noting: Southwood 1966, Hellawell 1978, CONCAWE 1982, Gilbertson et al. 1985, Moore & Chapman 1985, Chalmers & Parker 1986. Other references with information about recording and sampling equipment are noted appropriately in other parts of this book. There is also a range of specialized equipment designed for pollution monitoring, details of which can be found in specialized texts (eg. MARC Reports, see Appendix II) or more general texts such as Gilbertson et al. (1985).

Remote sensing and Geographical Information Systems (GIS)

Remote sensing techniques range from satellite imagery (used to monitor global environmental processes) and aerial photography (used to monitor salt marshes (Baker & Wolf 1987) and other ecosystems). The use of photography in monitoring is not restricted to aerial photography of landscapes; underwater stereophotogrammetry has been successfully used in monitoring benthic organisms and time-lapse video recording has been successfully used in monitoring behaviour of invertebrates in agricultural ecosystems (Halsall & Wratten 1988). One example of the use of satellite imagery for monitoring a salt marsh ecosystem is given on p. 172.

Interesting and recent advances in remote sensing may prove to be of considerable use for counts of wildlife. An exploratory project involving the Royal Society for the Protection of Birds and the National Remote Sensing Centre at Farnborough, England, is currently under way to explore the possibility of counting estuarine bird species with the help of satellite images from LANDSAT. The Nature Conservancy Council has also been exploring remote sensing techniques for monitoring moorland birds and their habitats, monitoring salt marshes and rare plant species in remote habitats. The first 30 000 records of the NCC's ancient woodland site databases have been transferred to GIS and this has resulted in a unique map of Britain's ancient woodland. See Chapter 9 for more details about uses of remote sensing and Geographical Information Systems – GIS.

The use of remote sensing techniques to monitor movements of animals is not new. For example, for many years now some very sophisticated radio transmitters have been used to study animal behaviour and monitor the movements of animals (Kenward 1987). Some biotelemetric studies have important applied aspects, particularly with regard to the provision of data for the conservation of various species and also for the design of management strategies to enable better coexistence of man and nature. One organization which aims to do exactly this is the Gallmann Memorial Foundation based in Kenya (see Appendix II) and more recently this Foundation together with the World Wide Fund for Nature has supported ecological monitoring of the Black Rhinoceros (*Diceros bicornis*), a species of rhinoceros which is both secretive and very much endangered (see Chapter 16 for comments on low genetic variation in rhinoceros).

The use of satellites to monitor the activties and movements of animals is, however, a relatively new venture and to date the movements of several species of animals have been monitored with the help of satellites. For example, Loffler & Margules (1980) have been able to monitor the burrowing and mound building of Australia's Hairy-nosed Wombats (*Lasiorhinus latifrons*) with digitally enhanced colour LANDSAT imagery. This species of wombat occurs in large numbers in the Nullarbor Plain of South Australia and it has been the subject of considerable controversy between farmers and conservationists. The authors claimed that the enhanced imagery provides a tool to map the approximate distribution of wombat colonies, thus providing useful data which are necessary for effective wildlife management.

The quite dramatic distances travelled by the Wandering Albatross (*Diomedea exulans*) have recently been revealed with the use of satellite telemetry (Jouventin & Weimerskirch 1990). Distances of between 3600 and 15 000 km were covered in a single foraging excursion with distances of up to 900 km per day. Such detailed data may prove critical for the conservation of this and other endangered, closely related species.

References

Baker, J. M. & Wolf, W. J. (1987). *Biological Surveys of Estuaries and Coasts.* Cambridge, Cambridge University Press.

Bloomfield, P. (1976). *Fourier Analysis of Time Series: an Introduction.* New York, London, John Wiley & Sons.

Bowles, M. L., Hess, W. J., DeMauro, M. M. & Hiebert, R. D. (1986). Endangered Plant Inventory and Monitoring Strategies at Indiana Dunes National Lakeshore. *Natural Areas Journal*, **6**, 18–26.

Bradshaw, M. E. (1981). Monitoring grassland plants in Upper Teasdale, England. In *The Biological Aspects of Rare Plant Conservation*, ed. H. Synge, pp. 241–57. Chichester, John Wiley.

Caughley, G. (1977). *Analysis of Vertebrate Populations*. Chichester, Wiley.

Caughley, G., Sinclair, R. & Scott-Kemmis, D. (1976). Experiments in aerial survey. *The Journal of Wildlife Management*, **40**, 290–300.

Chalmers, N. & Parker, P. (1986). *The OU Project Guide. Fieldwork and Statistics for Ecological Projects*. Field Studies Council, Taunton.

CONCAWE (1982). *Ecological Monitoring of Aqueous Effluents from Petroleum Refineries*. CONCAWE Report No. 8/82, Den Haag.

Digby, P. G. N. & Kempton, R. A. (1987). *Multivariate Analysis of Ecological Communities*. London, New York, Chapman & Hall.

Fisher, R. A., Corbet, A. S. & Williams, C. B. (1943). The relation between the number of species and the number of individuals in a random sample of an animal population. *Journal of Animal Ecology*, **12**, 42–58.

Gilbertson, D. D., Kent, M. & Pyatt, F. B. (1985). *Practical Ecology for Geography and Biology. Survey, Mapping and Data Analysis*. London, Hutchinson.

Govaere, J. C. R., Van Damme, D., Heip, C. & De Coninck, L. A. P. (1980). Benthic communities in the southern bight of the North Sea and their use in ecological monitoring. *Helgolander Meeresunters*, **33**, 507–21.

Gregory, P. T. (1978). Feeding habits and diet overlap of three species of garter snakes (*Thamnophis*) on Vancouver Island. *Canadian Journal of Zoology*, **56**, 1967–74.

Hales, D. C. (1962). Stream bottom sampling as a research tool. *Proceedings of the Utah Academy*, **39**, 84–91.

Halsall, B. B. & Wratten, S. D. (1988). The efficiency of pitfall trapping for polyphagous predatory Carabidae. *Ecological Entomology*, **13**, 293–9.

Hellawell, J. M. (1978). *Biological Surveillance of Rivers. A Biological Monitoring Handbook*. NERC, Water Research Centre, Regional Water Authorities.

Hiscock, K. (ed.) (1985). *Rocky Shore Survey and Monitoring Workshop, 1–4 May 1984*. London, BP International.

Hutchinson, G. E. (1959). Homage to Santa Rosalia, or why are there so many kinds of animals? *American Naturalist*, **93**, 145–59.

Jouventin, P. & Weimerskirch, H. (1990). Satellite tracking of Wandering Albatrosses. *Nature*, **343**, 746–8.

Kays, S. & Harper, J. L. (1974). The regulation of plant and tiller density in a grass sward. *Journal of Ecology*, **62**, 97–105.

Kenward, R. E. (1987). *Wildlife Radio Tagging: Equipment, Field Techniques and Data Analysis*. Academic Press.

Kershaw, K. A. (1964, 1973). *Quantitative and Dynamic Plant Ecology*. London, Edward Arnold.

Kirby, K. J. (1988). *Woodland Survey Handbook*. Research and Survey in Nature Conservation, No. 11, NCC, Peterborough.

Lewis, T. & Taylor, L. R. (1967). *Introduction to Experimental Ecology, a Student's Guide to Fieldwork and Analysis*. London, Academic Press.

Little, A. & Hiscock, K. (1989). Rocky-shore monitoring. In *Ecological Impacts of the Oil Industry*, ed. B. Dicks, pp. 9–35. Chichester, John Wiley.

Loffler, E. & Margules, C. (1980). Wombats detected from space. *Remote Sensing of Environment*, **9**, 47–56.

Means, D. B. & Simberloff, D. (1987). The peninsula effect: habitat correlated species decline in Florida's herpetofauna. *Journal of Biogeography*, **14**, 551–68.

Millard, S. P. & Lettenmaier, D. P. (1986). Optimal design of biological sampling programs using the analysis of variance. *Estuarine, Coastal and Shelf Science*, **22**, 637–56.

Mitchell, R. & Probert, P. K. (1981). *Severn Tidal Power, The Natural Environment*, NCC, Peterborough.

Moore, P. D. & Chapman, S. D. (1985). *Methods in Plant Ecology*, 2nd ed., Oxford, Blackwell Scientific.

Ovington, J. D. (1965). *Woodlands*. London, English University Press.

Pianka, E. R. (1969). Sympatry of desert lizards (*Ctenotus*) in Western Australia. *Ecology*, **50**, 1012–30.

Preston, F. W. (1948). The commonness and rarity of species. *Ecology*, **29**, 254–83.

Preston, F. W. (1962). The canonical distribution of commonness and rarity: part 1. *Ecology*, **43**, 185–215.

Simpson, G. G. (1964). Species density of North American recent mammals. *Systematic Zoology*, **12**, 57–73.

Southwood, T. R. E. (1966). *Ecological Methods, with Particular Reference to the Study of Insect Populations*. London, Methuen.

Spellerberg, I. F. & Pritchard, A. J. (1984). Ecology, ecosystem management and biology teaching. *Biology and Human Welfare*, No. 10. Commission for Biological Education of the International Union of Biological sciences and UNESCO.

Steele, B. B., Bayn, R. L. & Grant, C. V. (1984). Environmental monitoring using populations of birds and small mammals: analysis of sampling effort. *Biological Conservation*, **30**, 157–72.

Taylor, L. R. (1961). Aggregation, variance and the mean. *Nature*, **189**, 732–5.

Vernberg, F. J. & Vernberg, W. B. (1970). *The Animal and the Environment*. New York, Holt, Rinehart & Winston.

Vickerman, G. P. (1985). Sampling plans for beneficial arthropods in cereals. *Aspects of Applied Biology*, **10**, 191–8.

Warwick, R. M. & Uncles, R. J. (1980). *The Relationship Between the Sub-tidal Benthic Fauna and the Types of Bottom Sediment in the Bristol Channel and Severn Estuary*. Institute for Marine Environmental Research, NERC.

5

Biological indicators

Introduction

ONE DEFINITION of an indicator, as given in the *Oxford Dictionary of Current English*, is 'a device indicating the condition of a machine etc.' Replace 'device' with 'species' and replace 'machine' with 'the environment' and we have a reasonable definition of an indicator species. In a much wider sense, there are indicators of the state of the environment, both natural and man-made environments (examples in Chapter 8).

Physical disturbance and changes in environmental variables such as temperature or salinity will result in a change in the species composition of the biotic community. Such community changes can usefully be used in monitoring current states of the environment and also in a predictive sense in relation to environmental assessments (see Chapter 15). The Environmental Protection Agency in the USA, the United Nations Economic and Social Council, the United Nations Environment Programme and other agencies have drawn up lists of indicators in the context of indicating environmental quality. Amongst these lists of indicators (of the states of the environment) are biological variables which could be used periodically to assess pressures on the environment. The operation of these environmental indicator systems is, however, in its infancy according to a review by Vos *et al.* (1985).

By way of contrast there has been much detailed research on the use of biological indicators for the detection of pollution and specific pollutants. Biological material and indicator species used for monitoring of pollutants are many and varied, ranging from cells, tissues and organs to whole organisms (Table 5.1) including unicellular organisms, plants and animals (Thomas, Goldstein & Wilcox 1973, Burton 1986).

Table 5.1. *Examples of some reports from the 'popular' and 'scientific' literature about the use of indicator species for monitoring environmental contaminants*

Taxa	Reference
'Lower' plants	Burton, M. A. S. (1986). *Biological Monitoring of Environmental Contaminants*, MARC, London.
Lichens	Skye, E. (1979). *Annual Reviews of Phytopathology*, **17**, 325–41.
'Higher' plants	Manning, W. J. R. & Feder, W. A. (1980). *Biomonitoring Air Pollutants with Plants*. Applied Science Pub., London.
Coelenterates	Hanna, R. G. & Muir, G. L. (1990). *Environmental Monitoring and Assessment*, **14**, 211–22.
Aquatic invertebrates	Thomas, W. A., Goldstein, G. & Wilcox, W. H. (1973). *Biological Indicators of Environmental Quality*. Ann Arbor Science Pub., Michigan.
Terrestrial invertebrates	Allred, D. M. (1975). *Great Basin Naturalist*, **35**, 405–6.Ramsay, G. W. (1989). In Craig (1989), pp. 240–7 (see references).
Earthworms	Czarnowska, K. & Jopkiewicz, K. (1980). *Bioindication 3, Univ. Halle-Witenberg, Wiss. Beitr.*, **20**, 69–74.
Bees	Samiullah, Y. (1986). *Bee Craft*, **68**, 5–11.
Fish	Gruber, D. S. & Diamond, J. M. (1988). *Automated Biomonitoring, Living Sensors as Environmental Monitors*. Chichester, Ellis Horwood.
Snakes	Bauerle, B. (1975). *Copeia*, **1975**, 366–8
Birds	Ratcliffe, D. A. (1980). *The Peregrine Falcon*. Calton, T. & A. D. Poyser.
Marine birds	Batty, L. (1989). *Biologist* (J. of Inst. Biol.), **36**, 151–4.

Plant and animal indicators

Presence and absence

On p. 14 there was reference to Hardy's description of plankton monitoring. In the same very readable book (Hardy 1956) there is a nice example of how the presence and absence of a plankton species can be an indicator of environmental condition. Hardy described the distribution of Arrow Worms (*Sagitta setosa* and *Saggita elegans*) and noted that coastal water can be distinguished from the more oceanic water by the presence of these species of *Sagitta*; the former being found in coastal water and the latter in oceanic water.

In general, the presence of every plant and animal (and its condition) is a measure of the conditions under which it is existing or existed previously (see p. 9 for comments on indicators of past conditions). For example, occurrence of Common Stinging Nettles (*Urtica dioica*) is an indication of possible high levels of nitrogen in the soil, the appearance of Rosebay Willow Herb (*Chamaenerion angustifolium*) indicates disturbed soil or some kind of perturbation. Throughout history, different cultures have known that the presence of certain species, especially plant species, indicated certain conditions (or that certain conditions were required for growth of certain plants). For example West Africans have long recognized good soil by the presence of the Gau Tree (*Acacia albida*), the Gaya Grass (*Andropogon gayanus*) and also the Roan Antelope (*Hippotragus equinus*) (Zonneveld 1983). Plant indicators have had an interesting application in the location of over 70 different minerals. For example, the presence of a species of basil (*Ocimum homblei*) in Zimbabwe is an indication of a high copper content in the soil. The use of indicator plants has also been widely used in prospecting for gold (Erdman & Olson 1985). There are many species of animals, the presence of which can be used as indicators of certain environmental condition, such as high organic load. The insect group Ephemeroptera (Mayflies) which have aquatic larvae contains species which, with very few exceptions, are intolerant to organic enrichment, and so they have been incorporated into programmes monitoring water quality (see Table 8.3).

In practice, the use of presence or absence of organisms as indicators of environmental conditions requires some caution because of the variability inherent between populations. Genetic differentiation amongst some plant species, for example, can lead to some populations being commonly associated with base soils but in other parts of the species' range, populations may tolerate slightly acid conditions.

Behaviour and physiology

The behaviour and physiology of some animals have been useful in monitoring the quality of the environment. At a simple level, and an example familiar to us all, is the dawn chorus. A survey of those species of birds taking part in the dawn chorus at different locations in an urban area will soon show that near disturbed and polluted areas, fewer bird species take part in the chorus; a phenomenon which may remind us of Carson's *Silent Spring* (Table 2.1). Still with birds, the behaviour of a caged canary succumbing to dangerous gases has long been used by miners to detect methane; in this instance the canary is a biological early warning system (BEWS).

The behaviour and respiratory physiology of several aquatic organisms including fish have successfully been used to monitor water quality. For example, the respiratory and cardiac activity of various species of trout (*Salmo*) and other fish species have been used in sophisticated, automated BEWS (see, for example, Evans & Wallwork 1988). Diamond, Collins & Gruber (1988) of Biological Monitoring Inc. have reviewed research and development of BEWS and although supporting the use of fish and other organisms to monitor water quality and to act as early warning systems, they have suggested that the additional use of chemical monitoring for basic water quality parameters can help to identify causes of changes in water quality. In other words they support the combined use of biological and chemical monotoring.

One particularly interesting fish species used as a 'bio-sensor' is the Elephant-nosed Mormyrid (*Gnathonemus petersi*). This species navigates and communicates by emitting sequences of pulses from an electric organ on its tail. So sensitive and reliable is this behaviour that a research and development company (Aztec Environmental Control Ltd) has devised a sophisticated pollution monitoring system based on the computer-controlled detection of the pulses emitted by Mormyrids. However, not all individual fish will exhibit the same response and therefore the use of this fish as an indicator has to be based on the mean results from a group of fish. Almost certainly during the next few years we will see some exciting developments and applications in which these 'sensitive' species will be used in biological and ecological monitoring of water quality.

Microevolution

One classic example of an indicator of the extent of pollution has been the spread of melanic forms of the Peppered Moth (*Biston betularia*) throughout polluted areas of Britain and Europe. Particularly striking was the spread but then later decline of melanic forms after the Clean Air Act in Britain. This decline in melanism coincided with a period of increasing species richness of lichens on trees (Berry 1990, Brakefield 1990).

Community indicators

Populations of animals and plants occur in communities and therefore the species indicator concept can be extended to communities of indicator species. Different soils (for example serpentine soils, chalk soils and acid soils) all support indicator plant communities. The characteristic flora of serpentine soils, which are low in Ca and high in Mg, is a good example of a plant indicator community (Whittaker 1954). In North American serpentine soils, the vegetation is usually sparse and stunted and includes narrowly endemic species (e.g. *Quercus durata, Caeonothus jepsoni, Garrya congdoni* and *Cupressus sargentii*) which make a striking contrast with vegetation on other nearby soils. Another example is heathlands which are found on oligotrophic, acid soils and the indicator heathland plants are the low-growing, dwarf ericoid shrubs. Use of plant associations as an indicator of soil chemical composition and even depth of ground water has been undertaken in the USSR (Khudyakov, 1965) where water of 'best quality' was discovered under associations with dominance of *Alopecurus pratensis, Agropyron pectiniforme* and *Stipa capillata*.

Indicators of pollution

It has long been known that heavy metals and organochlorides penetrate ecosystems and, as a result, some oganisms will accumulate pollutants in varying amounts. For example, in polluted parts of the River Thames, the mollusc *Anodonta* spp. has been found to have twenty times the level of cadmium compared to the same species from the River Test near Southampton Water (Leatherland & Burton 1974). Although bioaccumulation occurs in a wide variety of taxanomic groups, it does not necessarily follow that the source of pollution is near those organisms in which the pollutants have accumulated. An extreme but worrying example was the discovery of DDE (a derivative of DDT) in bodies of Penguins in the Antarctic, thousands of miles from any use of agrochemicals. Nevertheless, a wide range of species have been used as pollution indicator species (Table 5.1) and many more species, especially marine species, have recently been tested for their use as indicator species. The Plymouth Marine Laboratory (NERC) is one of the research organizations which has undertaken much research on this important topic.

Obviously there are many uses of the term indicator species and this varied use provides a basis for distinguishing between the types of indicators as follows.

1. Sentinels: sensitive organisms introduced into the environment, for example, as early warning devices (canaries in coal mines) or to delimit the effect of an effluent.
2. Detectors: species occurring naturally in the area of interest and which may show a measurable response to environmental change, e.g. changes in behaviour, mortality, age-class structure etc.

3. Exploiters: species whose presence indicates the probability of disturbance or pollution. They are often abundant in polluted areas because of lack of competition from eliminated species.
4. Accumulators: organisms that take up and accumulate chemicals in measurable quantities.
5. Bioassay organisms: selected organisms used as a laboratory reagent to detect the presence and/or concentration of pollutants, or to rank pollutants in order of toxicity.

In this book we are concerned mainly with changes in ecological systems and methods of detecting and monitoring changes rather than the use of organisms to detect levels and extent of pollution. The latter topic, which makes use of accumulator types of indicators, is verging on another large area, environmental toxicology and ecotoxoicology. In this section we consider mainly the detector-type indicator with some reference to the exploiter- and accumulator-type indicators. Examples of exploiter-type indicators are given in Chapters 8 and 12, the latter with reference to monitoring water quality.

Detectors and exploiters

A number of authors (e.g. Butler *et al.* 1971, Buikema, Niederlehner & Cairns 1982) have reviewed and suggested desirable properties of detector and exploiter types of indicator species in connection with monitoring pollution; these properties include the following.

1. The organism should have narrow tolerances to environmental variables, for example stenothermal, stenohaline as opposed to erythermal and euryhaline.
2. The organism should be sedentary or have a limited dispersal.
3. Easy to sample and therefore presumably common would be an advantage.
4. Accumulation of pollutants should occur without killing the organism (that is unless mortality is used as the variable).
5. Preferably the organism should be long-lived so that different age-classes can be sampled.

As plants are sedentary and many are easy to sample, it is not surprising that they have had wide use as a detector and exploiter type of indicator. Responses of plant indicator species may take the form of changes in distribution or tissue damage. Symptoms of chronic injury include premature senescence and bronzing or chlorosis. However, results should be interpreted carefully because some injuries thought to be caused by pollutants may have been caused by disease, insects or environmental stress such as drought. The history of extensive studies has culminated in some useful illustrative works which help to identify visible pollutant effects on plants (see, for example, van Haut & Stratmann 1970).

Enterprising work on the use of diatoms and the relative abundance of diatom species has explored their use in monitoring river and stream water quality (Cairns 1981). Sampling the diatoms can be undertaken by using artificial substrata such as glass slides but sampling needs to be very frequent because some species of diatoms respond very quickly to pollution. Continuous monitoring of diatom species would seem impossible but the use of laser holography for identification of diatoms is an exciting technique which has been researched by Cairns et al. (1979, 1982). Three-dimensional holographs are used to record patterns on the diatoms and thus provide a basis for identification; information is then stored on a computer.

Lichens and mosses are long-lived organisms and their sensitivity to airborne pollutants is well known. In the case of lichens, pollutants such as SO_2 affect the algae component of the lichen and thus the symbiotic relationship between algae and fungus breaks down. The use of lichens as bioindicators has been so well documented over the last 130 years (Hawksworth & Rose 1976) that there are now many 'easy to follow guides' for use of lichens in monitoring air pollution (see, for example, James 1982). Philip Harris Biological Ltd (se Appendix II) have a good introductory pack for lichen surveys and air pollution studies.

It is the sensitivity of lichens as well as their long life span which makes them useful as indicator species, especially for sulphur dioxide levels. For example, Hawksworth & Rose (1976) have developed a method of estimating mean winter levels of atmospheric sulphur dioxide based on the presence of certain indicator lichen taxa. Using a scale of 0 to 10 (highest and lowest concentrations of SO_2) it is possible to prepare maps with zones of pollution based on the lichen species composition in each zone. One limitation of lichens as indicator species is that they are rather slow to respond to changes in levels of sulphur dioxide, that is the effects take place over years rather than in weeks or days.

In North America and in Europe, the use of lichens as 'detector', 'exploiter' and also 'accumulator' type of indicator of air pollution has been extensively described (see, for example, Le Blanc & Sloover 1970, Johnsen & Sochting 1973). Many questions remain unanswered about the physiological basis of effects of SO_2 on lichens and although there may be confounding factors such as microclimate, droughts and the buffering effects caused by the substrata (good reviews by Ferry, Baddeley & Hawksworth 1973, Skye 1979), lichens have been used successfully to

Table 5.2. *Increase in lichen abundance in North-west London based on a survey of 29 sites. All species were previously extinct or very rare*

Lichen species	Sites now present	Percentage of 29 sites	Mean and range of abundance[a]
Evernia prunastri	8	27.6	1.6 (1–4)
Hypogymnia physodes	16	55.2	1.8 (1–3)
Parmelia caperata	1	3.4	1
Parmelia subaurifera	7	24.1	1.7 (1–4)
Parmelia sulcata	13	44.8	1.8 (1–5)
Usnea subfloridana	2	6.9	2 and 4

[a]Abundance is scored on a 1–5 scale with 5 maximum.
From Rose and Hawksworth (1981).

monitor air pollution either as an accumulator indicator (see below report of work by Pilegaard 1978) or as a detector-type indicator. There are many studies which describe changes in lichen ecology as a result of increasing acidity of the substratum and conversely an increase in abundance of lichens with decreasing pollutants. Rose & Hawksworth (1981), for instance, have shown that several species of lichen which were extinct or very rare in Greater London around 1970 have since extended their ranges considerably and become more abundant (Table 5.2). From about 1962 onwards the levels of SO_2 recorded at six London recording stations have decreased and this change in level of pollutants would appear to be the main reason for the lichen recolonization.

Changes in the epiphytic cover on trees, with special note being made of the lichens, has been used in a useful manner by Cook, Rigby & Seaward (1990) in a study of melanic frequencies in the Peppered Moth *Biston betularia*. Indeed, the method used (a method developed by Lees, Creed & Duckett 1973) and their method of data collected could usefully form part of a monitoring programme at a long-term ecological monitoring site or permanent vegetation plot (see Tables 3.1 and 9.2). In their study, epiphytic cover at 1.5 m from ground level on oaks was scored as a fraction of the circumference occupied by bare bark, pleurococcoid algae, crustose lichens, foliose lichens and bryophytes. The height of 1.5 m is a fairly standard height and is used in other studies such as population studies of trees where diameter of the tree at 1.5 m would be recorded. High lichen species richness and cover was found at some sites well away from sources of pollution but as might be expected there was a lack of uniformity in the results.

There have been some fascinating applications of plant indicator species in the detection of ancient woodlands, classification of woodlands and in the ageing of hedges. Prior to recording the distribution and status of ancient woodlands and subsequent surveillance, there has to be a process of woodland classification and ancient woodland identification. The different rates at which various flowering plants and ferns are able to colonize new woodlands is the basis of ancient woodland indicators (Peterken 1974, Bunce 1982) and it is, for example, possible to identify a number of woodland vascular plant species which are strongly associated with ancient woodland. Lists of typical indicator species of ancient woodlands and recent secondary woodlands have now been established for surveys which, when completed, could provide baseline information for future monitoring programmes. One example of part of a list of woodland vascular plants associated with ancient woodland in the south of England is shown in Table 5.3.

Historical recording of hedgerows became popular following Hooper's (1970) publication of a method for dating hedgerows. That method has since contributed to some landscape assessments and surveillance of those landscapes. Hooper's method was very simple and was based on the idea that a 30 yd (27.4 m) length of hedge contains approximately one woody species for each 100 years of its existence (Pollard, Hooper & Moore 1974). Willmot's (1980) thorough research on dating hedges using woody species has since shown that less of the variation in number of woody species in hedges is due to the age of the hedge than has previously been implied. Previous management of the hedge, its location (whether next to a road or between fields) and other variables could all contribute to the number of woody species found. In addition, the fact that some hedges are remnants of woodlands or have been part of a woodland would affect the number of woody species found. Consequently Willmot (1980) concluded that the number of woody species as an indicator of hedge age should only be used for dating groups of hedges when a local relationship has been established empirically between age and number of species.

The use of whole live animals as detector indicators of air pollution is rare and this is perhaps because there are few suitable sedentary terrestrial species. One recent and interesting suggestion for monitoring and mapping air pollution involves the use of an Oribatid Mite, common in some orchards (Andre, Bolly & Lebrun 1982). The environmental biology of the mite *Humerobates rostrolamellatus* has been well studied and it is known to be sensitive to SO_2. The response time of this species to air pollution is shorter than that for lichens and the doubts about lichens because of confounding factors such as drought do not apply to these mites. The method used by Andre *et al.* was quite simple; adult mites collected from orchards in the country were placed in vials which were then left at 24

Table 5.3. *Ancient woodland vascular plants in woodland in the south of England. The following 36 species (from a list of 100 species) are thought to be most indicative of a long continuity and are therefore good indicators of an ancient woodland*

Adoxa moschatellina	Moschatel, Townhall clock
Allium ursinum	Wild Garlic
Anemone nemorosa	Wood Anemone
Blechnum spicant	Hard-fern
Carex laevigata	Smooth Sedge
Carex strigosa	
Convallaria majalis	Lily-of-the-valley
Daphne laureola	Spurge Laurel
Dryopteris carthusiana	Narrow Buckler-fern
Dryopteris pseudomas	
Epipactis purpurata	Violet Helloborine
Euphorbia amygdaloides	Wood Spurge
Galium adoratum	Sweet Woodruff
Helleborus viridis	Green Hellebore
Hordelymus europeaus	Wood Barley
Hypericum androsaemum	Tutsan
Lathrea squamaria	Toothwort
Luzula forsteri	Forster's Woodrush
Luzula sylvatica	Greater Woodrush
Melampyrum pratense	Common Cow-wheat
Milium effusum	Wood Millet
Oxalis acetosella	Wood Sorrel
Paris quadrifolia	Herb Paris
Platanthera chlorantha	Greater Butterfly Orchid
Polygonatum multiflorum	Solomon's Seal
Polystichum setiferum	Soft Shield-fern
Populus tremula	Aspen
Quercus petraea	Sessile Oak
Ranunculus auricomus	Goldilocks
Sanicula europaea	Sanicle
Sorbus torminalis	Wild Service Tree
Thelypteris oreopteris	Mountain Fern
Ulmus glabra	Wych Elm
Vaccinium myrtillus	Bilberry
Veronica montana	Wood Speedwell
Viola reichenbachiana	Pale Food Violet

Source: Nature Conservancy Council Ancient Woodland Vascular Plant List.

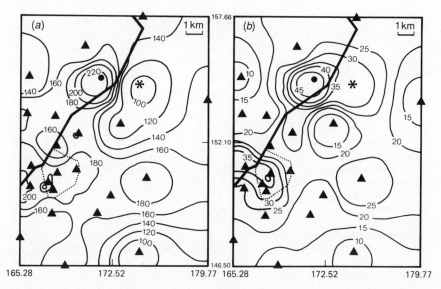

Fig. 5.1.(a) The SO$_2$ mean level contour map (micrograms per cubic metre).
(b) Mortality contour map (in per cent) for the Oribatid Mite *Humerobates rostrolamellatus*. The heavy line represents the Willebroek Canal and the dotted line represents the pentagonal inner loop of Brussels. The stations are indicated by triangles except sites 7 and 8 which are, respectively, represented by a circle and an asterisk. Lambert coordinates (i.e. true directions at each point) are specified in the margin. Redrawn from Andre *et al.* (1982).

pollution monitoring sites in Brussels. The mites were retrieved after a few weeks of exposure and the mortality was then used to construct the map in Fig. 5.1. Two high centres of mortality can be seen, one in the centre of Brussels and one near an industrial park on the Willebroek Canal. The use of this mite as a bioindicator seems quite convincing although further research such as on synergistic effects needs to be undertaken.

The use of detector and exploiter indicators for monitoring water quality has been described for many years but in early monitoring studies such as that undertaken by Gaufin & Tarzwell (1952) in Ohio, USA, it was realized that single invertebrate species have limited use as indicators. Their straightforward recording (Table 5.4) gave a classic result showing the differential effect on the biota caused by various chemical and physical conditions. However, they realized that, although absence of certain species from formerly clean water suggests pollution, single species of organisms such as *Tubifex tubifex* or *Chironomus tentans* can't be used as indicators of pollution unless relative abundance is recorded. Although the move away from reliance on single species to communities began many

Table 5.4. *Aquatic invertebrates as indicators of stream pollution in Ohio, USA*

Indicator species	'Unpolluted' May	'Unpolluted' Aug	'Polluted' May	'Polluted' Aug	'Recovering' May	'Recovering' Aug
Diptera						
Chironomus flarus	P					
Chironomus tentans					P	A
Chironomuc flaricingula	P				P	
Chironomus quadripunctatum		P				
Chironomus sp. A		P				
Chironomus sp. B		P				
Pentaneura flavifrons	P	P				
Dictya sp.		P				
Anophcles punctipennis		P				
Culex pipiens			P	A	P	A
Eristalis sp.			A	A		
Stratiomyia sp.			P	P		
Nemotelus sp.	P					
Tabanus sp.	P		P	P	P	P
Brachydeutera argentata					P	
Simulium vittotum	P					
Hemerodromia sp.	P					
Pilaria sp.	P					
Tipula sp.	P					
Erioccra sp.	P					
Paradixa sp.	P					
Coleoptera						
Stenelmis crenata	P	P				
Stenelmis sp.		P				
Simsonia quadrinotata	P					
Bidessus sp. A		P				
Laccophilus sp.	P					P
Tropisternus lateralis						P
Tropisternus sp.	P	P	P	P	P	P
Laccobius sp.				P		P
Peltodytes sp.	P	P				P
Ephemeroptera						
Baetis cingulatus		P				
Baetis parvus	P					
Caenis sp.	P	P				
Callibaetis sp.		P				P
Stenonema femoratus	P	P				
Stenonema ohioense	P	P				
Trichoptera						
Brachycentrus americanus						
Cheumatopsyche sp.	P	P				
Hydropsyche betteni	P	P				

Table 5.4. (*cont.*)

Indicator species	'Unpolluted'		'Polluted'		'Recovering'	
	May	Aug	May	Aug	May	Aug
Chimarra obscura	P					
Colophilus shawnce	P					
Phyacophila lobifera	P					
Ochrotrichia sp.	P					
Hydroptila consimilis	P					
Hydroptila sp.	P					
Plecoptera						
Acroneuria evoluta			P			
Allocapnia viviparia	P					
Nemoura venosa	P					
Perlesia placida	P					
Isoperla minuta	P					
Odonata						
Plathemis sp.						P
Paltothemis sp.						P
Argia sp.	P					
Agrion sp.	P					
Enallagma sp.						P
Neuroptera						
Sialis sp.		P				
Hemiptera						
Belostoma sp.		P			P	P
Corixidae	P	P	P	P		
Gerris sp.	P	P				
Microvelia sp.		P				
Notonecta sp.						P
Ranatra sp.						P
Crustacea						
Asellus sp.	P	P				
Cambarus rusticus	P	P				
Hyalella sp.		P				
Mollusca						
Physa integra	P	P	P	P	P	A
Sphaerium solidulum	P	P				
Annelida						
Limnodrilus sp.			P	A	P	A
Tubifex sp.			P	A	P	A
Glossiphonia sp.						P

Abbreviations: A = abundant, P = present.
From Gaufin & Tarzwell (1952).

decades ago, detailed ecological studies underlying community indicators are relatively more recent.

We have previously noted that it is possible to undertake biological and ecological monitoring in retrospect (see p. 11). One very nice study using diatoms to indicate previous acidification of two Scottish lochs has been described by Flower & Battarbee (1983). There are few long-term ecological studies of acidification of lakes and it is of great interest therefore that these authors were able to use diatom analysis of core samples to show declines in pH during the past 130 years. Other studies of acidification of lakes have used an experimental approach and an example is given later in Chapter 9.

It is the change in the relative abundance of various species which form the basis of community indicators. Patrick (1972), for example, in her account of aquatic communities as indicators of pollution in North American natural streams, described how in many streams species richness at different trophic levels remains fairly similar throughout the year but species composition changes greatly. Thus, in natural streams there is no dominance by a single species: these circumstances change when pollution occurs and the change follows a pattern not unexpected with exploiter-type indicators becoming the dominant species. The first effects of an increase in organic load are to cause some species such as diatoms to become more common. The species composition of protozoans changes from well balanced to almost complete dominance by ciliates. Similarly there is a shift in the species composition of herbivores and carnivores and some of these are represented by large populations. It is these kinds of changes reflected in the indicator species that have been particularly useful in monitoring water pollution, by way of community indicators.

Accumulators

The effects of air pollutants on plants have been well researched and excellent, comprehensive reviews include Manning & Feder (1980) and Martin & Coughtrey (1982). Lichens have been used both as detector indicators and as accumulator indicators and Pilegaard's research, mentioned above, in an industrial area of Denmark is a good example of how lichens have been used as accumulator-type indicators to monitor airborne metals and sulphur dioxide. Using samples of the epiphytic lichen *Lecanora conizaeoides*, which is tolerant of SO_2, it was possible to show that concentrations of all nine heavy metals found in *L. conizaeoides* have step gradients with highest values nearest to the industrial area.

Hair, shells, bones and internal organs such as liver, kidneys and muscle from a wide range of animal indicator species have been used in biological monitoring of environmental contaminants (MARC 1985, Burton 1986).

Interesting examples of animal material being used as a bioindicator of environmental pollution include those examples using deer antlers. There are many collections of deer antlers throughout the world but without any baseline research those antlers could not reliably be used for monitoring environmental pollutants. However, thorough research by Jones & Samiulla (1985) has established that samples taken from deer of known ages can be used to monitor physiological and environmental metal concentrations despite possible variations in the concentrations of metals in different parts of the antlers. This research by Jones and colleagues could provide a baseline for further studies involving the use of antlers in collections representing a time-scale of several hundred years.

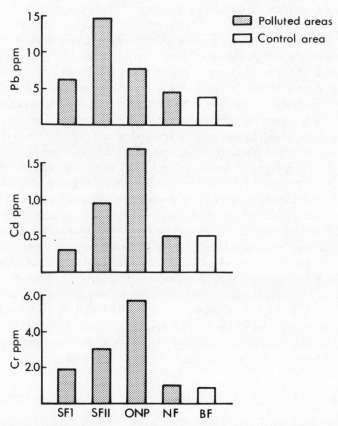

Fig. 5.2. Concentrations of lead, cadmium and chromium (ppm, dry weight) in Roe Deer antlers, SFI, Silesian forests 1938–50. SFII, Silesian forests 1951–73. ONP, Ojcow National Park. NF, Niepolomice Forest. BF, Bialowieza Forest. Redrawn from Sawicka-Kapusta (1979).

That deer antlers can be used successfully as bioindicators of environmental pollution has been confirmed in a study of Roe Deer in Polish forest by Sawicka-Kapusta (1979). For example, data for three metals from five groups of animals, one of which was a control, show very clearly that industrial pollutants do accumulate in these deer antlers (Fig. 5.2). Deer antlers would seem to be particularly useful as indicators of environmental pollution especially as the antlers represent about 130 days of growth and can be sampled annually without harm to the animals.

Although many animal species have been used as accumulator type indicators for monitoring levels of pollution in water (see review by Phillips 1980) there are a number of advantages and disadvantages compared to chemical monitoring. Assuming careful selection of the indicator organism for its known ability to accumulate particular pollutants under certain conditions, one advantage is that higher accumulations of the pollutant in the organism compared to concentration in the environment facilitate easier and cheaper analysis. Secondly, the temporal variation in the level of chemical pollutants in water can be overcome by the use of indicator organisms which allow a time-averaged index of pollutant availability. Phillips (1980) has also suggested that the use of indicator organisms enables availability of pollutants to be measured directly, thus avoiding the need to try and detect pollution levels from speciation of metals (occurrence of a metal in different forms).

The disadvantage of using accumulator indicator species in the detection of pollutants arises from the fact that a number of biotic and abiotic variables may affect the rate at which the pollutant is accumulated. Clearly both laboratory and field tests need to be undertaken so that the effects, if any, of extraneous parameters can be identified. The extraneous parameters can be grouped under three general headings. Firstly, abiotic parameters such as water temperature levels, salinity levels, pH and other parameters that may obscure or alter the availability of pollutants, either directly or via the behaviour of the organism. Secondly, biotic factors such as age, size, sex, stage in the sexual cycle, growth and diet, body lipids and behaviour may all affect the rate of uptake and rate of accumulation of metals or organochlorines. Thirdly, there may be synergistic effects resulting from the presence of more than one pollutant.

Status of biological indicators in monitoring programmes

There has been much research on the nature of biological indicators but a disappointing amount of research on the role of biological indicators in pollution monitoring programmes. The reason for this can be attributed mainly to the apparent cheaper methods of pollution detection made possible by sophisticated technology. Machines may be more reliable than biological organisms and it has to be admitted that care has to be taken when it comes to interpreting the physiology, behaviour or ecology of biological indicators. Cause and effect relations are never easy to confirm without good research.

Nevertheless, it is one thing to detect levels of pollutants and another thing to monitor the effects on organisms and ecosystems. All too often we have seen the effects of pollution on ecosystems but have had to spend much time and effort trying to establish the cause. There are many biological indicators which could successfully be used as not only as effective warning systems but also as cheap and reliable components of long-term pollution monitoring programmes. There is also much need for research on establishing baselines, such as has been undertaken for the RIVPACS system (see p. 234)

References

Andre, H. M., Bolly, C. & Lebrun, P. H. (1982). Monitoring and mapping air pollution through an animal indicator: a new and quick method. *Journal of Applied Ecology*, **19**, 107–11.

Berry, R. J. (1990). Industrial melanism and peppered moths (*Biston betularia* (L.)). *Biological Journal of the Linnean Society*, **39**, 301–22.

Le Blanc, F. & De Sloover, J. (1970). Relation between industrialization and the distribution and growth of epiphytic lichens and mosses in Montreal. *Canadian Journal of Botany*, **48**, 1485–96.

Brakefield, P. M. (1990). A decline of melanism in the peppered moth *Biston betularia* in The Netherlands. *Biological Journal of the Linnean Society*, **39**, 327–34.

Buikema, A. L., Niederlehner, B. R. & Cairns, J. (1982). Biological Monitoring part IV – toxicity testing. *Water Research*, **16**, 239–62.

Bunce, R. G. H. (1982). *A Field Key for Classifying British Woodland Vegetation*, Part 1. Cambridge, Institute of Terrestrial Ecology.

Burton, M. A. S. (1986). *Biological Monitoring of Environmental Contaminants*. London, MARC.

Butler, P. A., Andren, L., Bonde, G. J., Jernelov, A. & Reish, D. J. (1971). Monitoring organisms. In FAO *Technical Conference on Marine Pollution and its Effects of Living Resources and Fishing, Rome 1970. Suppl. 1. Report of the Seminar on Methods of Detection, Measurement and Monitoring of Pollutants in the Marine Environment*, pp. 101–12. FAO Fisheries Reports.

Cairns, J. (1981). The use of microcomputers in biology. *The Biologist*, **63**, 33–47.

Cairns, J., Almeida, S. P. & Fujii, H. (1982). Automated identification of diatoms. *BioScience*, **32**, 98–102.

Cairns, J., Kuhn, D. L. & Plafkin, J. L. (1979). Protozoan colonization of artificial substrates. *American Society for Testing and Materials, Special Technical Publication* **690**, 34–57.

Cook, L. M., Rigby, K. D. & Seaward, M. R. D. (1990). Melanic moths and changes in epiphytic vegetation in north-west England and north Wales. *Biological Journal of the Linnean Society*, **39**, 343–54.

Craig, B. (1989). *Proceedings of a Symposium on Environmental Monitoring in New Zealand with Emphasis on Protected Natural Areas.* Wellington, Department of Conservation.

Diamond, J., Collins, M. & Gruber, D. (1988). An overview of automated biomonitoring – past developments and future needs. In *Automated Biomonitoring: Living Sensors as Environmental Monitors*, ed. D. Gruber & J. Diamond, pp. 23–39. Chichester, Ellis Horwood.

Erdman, J. A. & Olson, J. C. (1985). The use of plants in prospecting for gold: a brief overview with a selected bibliography and topic index. *Journal of Geochemical Exploration*, **24**, 281–304.

Evans, G. P. & Wallwork, J. F. (1988). The WRc fish monitor and other biomonitoring methods. In *Automated Biomonitoring: Living Sensors as Environmental Monitors*, ed. D. Gruber & J. Diamond, pp. 75–90. Chichester, Ellis Horwood.

Ferry, B. W., Baddeley, M. S. & Hawksworth, D. L. (1973). *Air Pollution and Lichens*. London, The Athlone Press of the University of London.

Flower, R. J. & Battarbee, R. W. (1983). Diatom evidence for recent acidification of two Scottish lochs. *Nature*, **305**, 130–3.

Gaufin, A. R. & Tarzwell, C. M. (1952). Aquatic invertebrates as indicators of stream pollution. *Public Health Reports*, **67**, 57–64.

Hardy, A. C. (1956). *The Ocean Sea. Its Natural History: the World of Plankton*. London, Collins.

van Haut, H. & Stratmann, H. (1970). *Farbtafelatlas uber Schwefeldioxid-Wirkungen an Pflanzen*, Essen, W. Gerardet.

Hawksworth, D. L. & Rose, F. (1976). *Lichens as Pollution Monitors*. London, Edward Arnold.

Hooper, M. D. (1970). Dating hedges. *Area*, **4**, 63–5.

James, P. (1982). *Lichens and Air Pollution*. British Museum (Natural History) and BP Educational Services.

Johnsen, I. & Sochting, U. (1973). Influence of air pollution on the epiphytic vegetation and bark properties of deciduous trees in the Copenhagen area. *Oikos*, **24**, 344–51.

Jones, K. C. & Samiullah, Y. (1985). Deer antlers as pollution monitors. *Deer*, **6**, 253–5.

Khudyakov, I. I. (1965). The vegetation cover as an indicator of the chemical composition and depth of groundwaters. In *Plant Indicators of Soils, Rocks, and Subsurface Waters*, ed. A. G. Chikishev, pp. 16–18. Authorized translation from the Russian of the proceedings of the Conference on Indicational Geobotany, 1961. Consultant Bureau Enterprises, New York.

Leatherland, T. M. & Burton, J. D. (1974). The occurrence of some trace metals in coastal organisms with particular reference to the Solent area. *Journal of the Marine Biology Association of the United Kingdom*, **54**, 457–68.

Lees, D. R., Creed, E. R. & Duckett, J. G. (1973). Atmospheric pollution and industrial melanism. *Heredity*, **30**, 227–32.

Manning, W. J. & Feder, W. A. (1980). *Biomonitoring Air Pollutants with Plants*. Applied Science.

MARC (1985). *Historical Monitoring*. University of London, Monitoring and Assessment Research Centre.

Martin, M. H. & Coughtry, P. J. (1982). *Biological Monitoring of Heavy Metal Pollution, Land and Air.* Applied Science.

Patrick, R. (1972). Aquatic communities as indices of pollution. In *Indicators of Environmental Quality*, ed. W. A. Thomas, pp. 93–100. New York, London, Plenum Press.

Peterken, G. F. (1974). A method for assessing woodland flora for conservation using indicator species. *Biological Conservation*, 6, 239–45.

Phillips, D. J. H. (1980). *Quantitative Aquatic Biological Indicators. Their Use to Monitor Trace Metal and Organochlorine Pollution.* London, Applied Science Publishers Ltd.

Pilegaard, K. (1978). Airborne metals and SO_2 monitored by epiphytic lichens in an industrial area. *Environmental Pollution*, 17, 81–92.

Pollard, E., Hooper, M. D. & Moore, N. W. (1974). *Hedges.* London, Collins.

Rose, C. I. & Hawksworth, D. L. (1981). Lichen recolonization in London's cleaner air. *Nature*, 289, 289–92.

Sawicka-Kapusta, K. (1979). Roe deer antlers as bioindicators of environmental pollution in southern Poland. *Environmental Pollution*, 19, 283–93.

Skye, E. (1979). Lichens as biological indicators of air pollution. *Annual Reviews of Phytopathology*, 17, 325–41.

Thomas, W. A., Goldstein, G. & Wilcox, W. H. (1973). *Biological Indicators of Environmental Quality. A Bibliography of Abstracts.* Ann Arbor, Michigan, Ann Arbor Science Publications.

Vos, J. B., Feenstra, J. F., Boer, J. de., Braat, L. C. & Baalen, J. van (1985). *Indicators for the State of the Environment.* Institute for Environmental Studies, Amsterdam, Free University.

Whittaker, R. H. (1954). The ecology of serpentine soils. *Ecology*, 35, 258–88.

Willmot, A. (1980). The woody species of hedges with special reference to age in Church Broughton Parish, Derbyshire. *Journal of Ecology*, 68, 269–85.

Zonneveld, I. S. (1983). Principles of bio-indication. *Environmental Monitoring and Assessment*, 3, 207–17.

6

Diversity

Introduction

FOR MANY YEARS, environmental indices have been used to monitor pollution, changes in biotic communities and so-called 'environmental standards' or 'quality of the environment'. Environmental indices include those which are based on physical and chemical parameters, those based on biological parameters and also those based on perceived aesthetic qualities of the environment (see Chapter 8). Here we are dealing only with a few of the many indices with a biological basis. The aim of this and the following two chapters is to provide an introduction to a selection of biological indices, their calculation, uses and limitations. Applications of some of these indices in monitoring programmes are described in a later section.

Number of species, species composition and abundance

Perhaps the simplest variables which could be used in biological and ecological monitoring are the number of species, the species composition and the proportional abundance of species. The difference between the number of species, species composition and proportional abundance of species in a hypothetical community may be illustrated as shown in Fig. 6.1. Sample 1 or the 'baseline' contains four species. Samples 2, 4 and 5 also contain four species (the species richness is the same for each sample) but there are differences in species composition and total biomass. In sample 2 there are also four species but the species composition has changed. There is a difference between the relative abundance of each group of species in samples 1 and 4. The species richness of sample 3 is greater than found in the other samples. Sample 5 has a higher biomass but the species composition and proportional abundance of the species is the same as in sample 1.

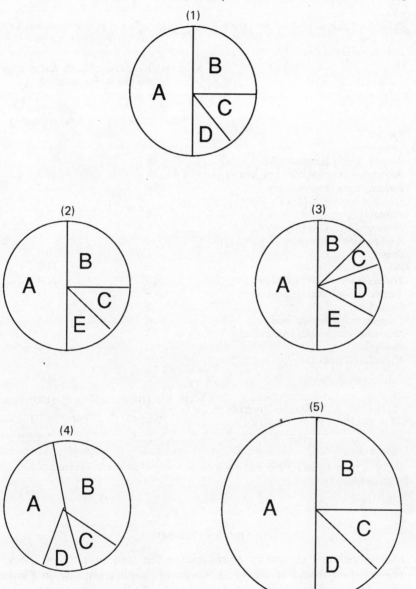

Fig. 6.1. Biotic community variables. Species indicated by letters A–E, abundance by the area of sector and biomass by area of the circle. Comparisons are made with 1. In 2 the species composition has changed, in 3 the species richness has increased, in 4 the relative abundance has changed and in 5 the biomass has increased. After Hellawell (1977).

Table 6.1. *'Sub-sample' of data from a study of the effect of people trampling on grassland*

Species	Species abundance (mean percentage cover per metre quadrat)	
	Site one 'high' trampling x	Site two 'low' trampling x
Agrostis tenuis (Common Bent-grass)	49	46.7
Poa annua (Annual Meadow Grass)	45	3.5
Festuca rubra (Red Fescue)	4	3.7
Plantago coronopus (Buck's Horn Plantain)	0	5.0
Plantago lanceolata (Ribwort)	0.5	1.3
Anthoxanthum odoratum (Sweet Vernal Grass)	1.0	0
Bellis perennis (Daisy)	0.5	11.3
Trifolium repens (White Clover)	0	21.2
Taraxacum officinale (Common Dandelion)	0	2.0
Prunella vulgaris (Self heal)	0	0.5
Leontodon autumnalis (Autumnal Hawkbit)	0	3.8
Hypochaeris radicata	0	1.2
Number of species	6	11

$$\text{Species diversity} = \frac{(\sum x)^2}{\sum x^2} = \frac{(100)^2}{49^2 + 45^2 \dots} \qquad \frac{(100)^2}{46.7^2 + 3.3^2 + \dots}$$

(Hill 1973) $= 2.25$ $\qquad$ 3.5

Modified from Jones (1979). Environmental Sciences Research Project, Southampton University.

Species richness

Species richness or species abundance is the total number of species present. Data from a study on the effects of people trampling on a lawn provide us with information (Table 6.1) with which to examine changes in species richness and also species composition. Abundance of the various plant species has been expressed as percentage cover in the sample areas. There is no doubt that the 'low trampling' site has a greater number of grassland species (higher species richness) than the 'high trampling' site. Both the species richness and the species composition in these samples could provide a basis for monitoring the effects of physical impacts on lawns. It is also useful to note that the 'high trampling' site is dominated by two species, *Agrostis tenuis* and *Poa annua* and that there is a marked difference in the abundance of *Poa annua*. The species diversity index in Table 6.1 is discussed below.

A sub-sample of a community from a collection of night-flying moths caught with a light-trap provides us with a further example with which to explore various features which might usefully be used in monitoring. Looking at Table 6.2 we see that the total number of species of moths (species richness) for each month varies between five and ten and was highest in July. The number of individuals for each species caught each month is used as a measure of abundance. Although total numbers of individuals of each species is an obvious way of expressing abundance or population size, alternative methods have been adopted for other taxonomic groups, such as percentage cover for grassland species (Table 6.1).

Species richness indices

Species richness can be expressed as the number of species in a sample or habitat, or could be expressed more usefully as species richness per unit area. There are also various simple species richness indices based on the total number of species and the total number of individuals in the sample or habitat. For example Menhinick's Index and Margalef's Index are simple measures of species richness and are expressed

$$D = \frac{S}{\sqrt{N}} \quad \text{(Menhinick 1964)}$$

$$D = \frac{S-1}{\log N} \quad \text{(Margalef 1951)}$$

where D is the index (Margalef's Index), S is the number of species and N is the total number of individuals. This is also a simple measure of mean population size and using the form $S-1$, rather than S, gives a value of zero if there is only one species. Despite its simplicity, this and similar indices are affected by sample size and sampling effort. Furthermore these indices could be misleading because they fail to take into account abundance patterns. For example, two samples (A, B) could have the same species richness (5) and same number of individuals (100) but the samples may differ in the proportional abundance of the species. This problem is considered in the following sections.

Table 6.2. *Species richness and species diversity of a 'sub-sample' of night flying moths collected with a light-trap at monthly intervals*

Family and species	Month			
	May	June	July	August
Amphipyrinae				
Apamea monoglypha (Dark Arches)	0	7	149	87
Apamea secalis (Common Rustic)	0	0	2	85
Meristis trigrammica (Treble Lines)	47	96	2	0
Arctiinae				
Spilosoma lutea (Buff Ermine)	4	44	31	0
Ennominae				
Biston betularis (Peppered)	22	36	92	5
Hadeninae				
Leucania impura	0	2	195	60
Lithosiinae				
Lithosia lurideola (Common Footman)	0	0	84	51
Noctuinae				
Agrotis exclamationis	1	144	255	38
Nuctua pronuba	0	13	32	87
Ochropleura plecta	25	41	24	3
Number of species	5	8	10	8
Number of individuals	99	383	866	416
Margalef's $\dfrac{S-1}{\log N}$	2.0	2.7	3.1	2.7
Simpson's or $\dfrac{1-\sum P_i^2}{\sum P_i^2}$	0.67	0.77	0.81	0.82
	3.0	4.3	5.3	5.6
Shannon–Wiener $-\sum P_i \log_e P_i$	1.2	1.6	1.8	1.8

P_i is the proportion of the ith species in the sample. For example, 47 *Meristis trigrammica* were caught in May. P_i in this case is $47/99 = 0.4747$, $P_i^2 = 0.225$.

Ecological diversity

In ecological research, diversity could refer to species diversity, habitat diversity or the diversity of resources in a niche. In other words ecological diversity embraces different kinds of diversity. One approach is to think in terms of alpha, beta and gamma diversity. Alpha diversity is the diversity of species within a particular habitat or community. Beta diversity is a measure of the rate and extent of change in species along a gradient from one habitat to another (or an expression of between-habitat diversity). Gamma diversity is dependent on both alpha diversity and beta diversity and is the diversity of species within a geographical area. Here we are concerned mainly with alpha diversity and the different ways of measuring and expressing that diversity.

Species diversity indices

An index of diversity can be based either on the number of species present and species composition without any measure of abundance (species richness index) or can be based on the species and abundance of species in a habitat or community (diversity index). In plant ecology studies, percentage cover of plant species in quadrats has had common use including long-term vegetation monitoring. For example, Belsky (1985) in her vegetation monitoring in the Serengeti National Park, Tanzania, found that of several variables used, percentage cover of species in permanently marked plots was the most useful for long-term studies. The data in Table 6.1 from the study of trampling have also been expressed as percentage cover but we can also see that the sample from the 'low trampling' site has greater species richness than the 'high trampling' site but how can we quantify this observation? Diversity can be based on percentage cover as a measure of abundance:

$$D = \frac{(\sum x)^2}{\sum x^2} \quad \text{(Hill 1973)}$$

where D is an index of diversity and x is a measure of abundance of the species, in this case percentage cover. This measure of diversity is one of many suggested by Hill (1973) and has been used in studies of the effects of trampling on vegetation (Liddle 1975). One example is shown in Table 6.1.

The sequential Comparison Index (Cairns et al. 1968) was devised as a simplified method for non-biologists to estimate relative differences in biological diversity and is based on the chance of any one individual being the same species as the previous individual in a sample. In the field, individual organisms are recorded at intervals along a transect. Random samples taken back to the laboratory are scored in turn as being the same species or different species to the preceding individual. Samples with many species each represented by few individuals will result in many changes and a high index value. For example, a hypothetical sample has three species (A, B, C) and when arranged in rows the arrangement is as follows:

$$CCAABAACBCAA$$

$$1 \quad 2 \quad 34 \quad 5678$$

After scoring one for the first individual, the number of changes of one species to another is eight and the total number of individual organisms is 13. A simple measure of the variety or diversity in this sample may be expressed as follows:

$$D \text{ (Diversity)} = x/N$$

where x is the number of changes and N is the number of individual specimens. In the above hypothetical example:

$$D \text{ (Diversity)} = 8/13 = 0.6$$

Williams & Dussart (1976) suggested that this index be based on a random collection of 200 organisms and that random numbers are needed in the laboratory once the individual has been classified to species. The reliability of this and other measures of diversity could be quantified by determining the diversity of the sample several times. In other words, what are the chances of the same measure of diversity being obtained if the sample were to be mixed then arranged in another set of rows? A few simple tests can usefully be undertaken on a random set of letters to test the reliability of Cairn's measure of diversity. In other words, it is useful to ask the question, how much variation could be expected to occur if the Sequential Comparison Index was calculated several times using the same specimens but with a complete mixing of the specimens between each calculation?

As explained in Chapter 4, communities contain many rare species and relatively few common species. The theories of abundance and the ease with which it is possible to calculate the total number of species and total number of individuals provides a basis for calculating diversity. For example, the logarithmic series for species abundance in a sample is fixed by two variables, the number of species and number of individuals (Southwood 1966):

$$S = \alpha \log_e(1 + N/\alpha)$$

where S is the total number of species, N is the total number of individuals and α is the diversity index.

The Alpha Diversity Index (Fisher, Corbet & Williams 1943, Williams 1964) is most simply read off from the nomogram (Fig. 6.2).

Species diversity and equitability or evenness

We can see how the individuals are distributed amongst the moth species (Table 6.2) and how the percentage cover of the plants is distributed amongst the various grassland species (Table 6.1) but is it possible to quantify the relative abundance and use this in the calculation of an index?

The way in which individuals are distributed among species is known as 'evenness' or 'equitability'. For example, consider an extreme hypothetical example of two samples, each of which has five species and a total of 100 individuals. In the first sample, each of the five species is represented by 20 individuals and in the second sample one species is represented by 96 individuals and the remaining four species by one individual each. The equitability is maximized in sample one and in the second sample there is an uneven distribution of individuals among the species. Species diversity is greatest in the first sample and lowest in the second sample. This equitability of evenness can be expressed in the form of one of several indices (heterogeneity indices) where a more equitable distribution among species will give higher index values. In this context, species diversity is a function of species richness and the evenness with which individuals are distributed amongst the species.

If the numbers of all species can be recorded (not just samples taken) then a suitable index of diversity is Brillouin's Index:

$$H = \frac{1}{N} \log_{10} \frac{N!}{N_1! \, N_2! \, N_3!} \quad \text{(Brillouin 1960)}$$

where H is the diversity index, N is the total number of individuals, N_1 is the number of individuals of species 1. This index has been derived for use where the total population can be measured (Pielou 1966) and is not appropriate where samples of the population have to used or where randomness of sampling cannot be guaranteed. It is an index which is not therefore subjected to limitations and error created by various sampling techniques. Brillouin's Index would be calculated as follows.

Assume the species A, B, C have the relative abundance of 4, 9, 5. Then

$$H = \frac{1}{18} \frac{(18 \times 17 \times 16 \dots \times 2)}{(4 \times 3 \times 2)(9 \times 8 \times 7 \dots \times 2)(5 \times 4 \times 3 \times 2)}$$

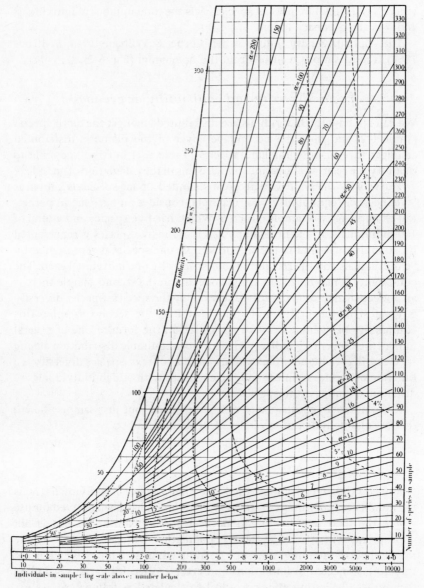

Fig. 6.2. Nomogram for determining the diversity index (α) for the number of species (S) and the number of individuals (N) in a random sample. From Fisher *et al.* (1943) with permission of Cambridge University Press.

One example of how this index can be used is described in a paper by Ben–Eliahu & Safriel (1982) where they made comparisons of polychaetes from tropical and temperate intertidal habitats. Although the aim was not to monitor changes in diversity of polychaetes, their survey does provide a most useful example of the kind of data used as a baseline for monitoring. They considered species richness and diversity 'H' as a function of sample size (Fig. 6.3) and found that the number of species and the diversity increased asymptotically with increased number of individuals. Further applications of these indices are described in Part C.

In circumstances where it is not practical or not possible to determine the total number of individuals then a random sample has to be used. There are many species diversity indices which are based on both the number of species and the proportional abundance of species. Two commonly used species diversity indices are Simpson's Index (Yules) and the Shannon–Wiener Index. The Simpson Index, D (Simpson 1949), is calculated as follows:

$$D = \sum_{i=1}^{s} P_i^2$$

where

$$(P_i^2) \sim \left(\frac{N_i}{N_T}\right)^2$$

where D is the index of diversity (a measure of the probability that two randomly sampled individuals belong to different species) and P_i is the relative abundance of the ith species (N_i is the number of the ith species, N_T is the total number of individuals). This index is sometimes used in either of the following forms:

$$D = \frac{1}{\sum P_i^2} \qquad D = 1 - \sum P_i^2$$

For example and with reference to Table 6.2 the diversity of the sample of moths collected during May would be calculated as follows:

$$D = \left(\frac{47}{99}\right)^2 + \left(\frac{4}{99}\right)^2 + \left(\frac{22}{99}\right)^2 \ldots$$

$$= 0.33$$

Or in the form

$$D = 1 - \sum P_i^2$$

$$= 0.67$$

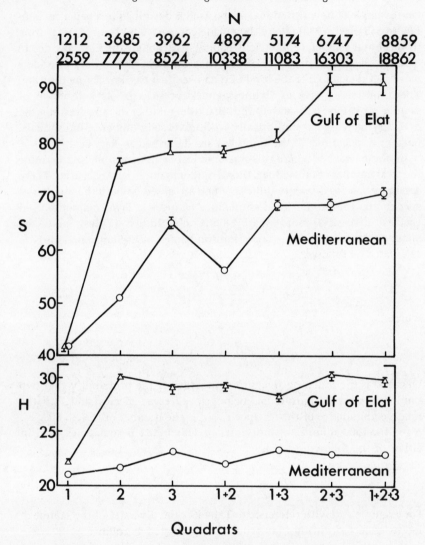

Fig. 6.3. Species richness (*S*, top) snd species diversity (*H*, bottom) as a function of sample size. N=number of individuals in samples (upper row, Gulf of Elat; lower row, Mediterranean). Circles (Mediterranean) and triangles (Gulf of Elat) stand for values obtained when all juveniles were allocated to their putative species. Vertical bars represent the ranges between minimal diversity obtained when unidentified juveniles were pooled with dominant congeners, and between maximal diversity by omitting the juveniles. Redrawn from Ben-Eliahu & Safriel (1982).

Predicting correctly the next species of the next individual to be collected provides a basis for another index. The Shannon–Wiener Index (see Pielou 1966 for comments about derivation) is calculated from the equation:

$$D = - \sum_{i=1}^{s} P_i (\log_e P_i)$$

where D is the species diversity index and P_i is the proportional abundance of the ith species in the same. The use of natural logs is usual because this gives information in binary digits ($\log_{10}$ of 100 is 2, $\log_e$ of 100 is 4.6, $\log_{10}$ of 1000 is 3, $\log_e$ of 1000 is 6.9).

With reference to data in Table 6.2, the Shannon–Wiener Diversity Index for the sample of moths collected during May would be calculated as follows:

$$D = -\sum \left(\frac{47}{99} \times \frac{47}{99} \log_e \right) \times \left(\frac{4}{99} \times \frac{4}{99} \log_e \right) \dots$$
$$= -\sum (0.47 \times -0.74) + (0.04 \times -3.2)$$
$$= -\sum (-0.35) + (-0.13)$$
$$= 1.2$$

There are some interesting values amongst the diversity indices for the samples of moths (Table 6.2). Eight species were caught in both June and August and so the species richness remained the same. The diversity index, if expressed as a simple relationship between the number of species and total number of individuals, was the sme for those two months. However, when we come to look at Simpson's Index of diversity and the Shannon–Wiener Index of diversity we see that the August sample has a greater value, indicating greatest evenness in the August sample compared to samples taken during other months. The lower diversity index values in June can be attributed to less evenness or less equitability of the sample, particularly with regard to the large number of individuals of one species *Agrotis exclamationis*.

Variation in species diversity

In simple terms, maximum diversity (equitability) exists if each individual belongs to a different species, minimum diversity exists if all individuals belong to one species. It does not necessarily follow that a higher species diversity is better than a lower species diversity nor does it follow that high diversity indices can be interpreted as being a reflection of high quality habitat. Some communities have a naturally low species diversity

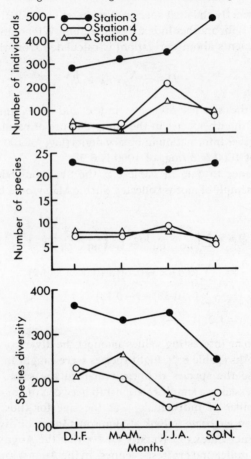

Fig. 6.4. Seasonal variations in the number of individuals, number of species and species diversity (Shannon–Wiener) at stations 3, 4 and 6 in Cane Creek Basin (Walker County, Alabama, USA) from February 1970 to January 1971. Redrawn from Dills & Rogers (1974).

and the meaning of 'quality' needs to be defined when using a diversity index to express some state of the habitat or environment. To complicate matters, not all species diversity indices are based on the idea of equitability and two different indices can give contradictory values. It is always useful to calculate the theoretical extremes of a diversity index with collection of simple, hypothetical data.

Temporal changes in species diversity must be expected, whether it be on a circadian, tidal, seasonal or annual basis. Dills & Rogers (1974) give a good example of seasonal variations in diversity (Shannon–Wiener) from their research on macro-invertebrate community structure affected by acid mine pollution in small streams in Alabama (Fig. 6.4). Changes in species diversity will also occur over time as a result of changes in temperature levels. This has been well investigated by Dennis, Patil & Rossi (1979) and an excellent example is that reported by Goldman in 1975 (Fig. 6.5) for species diversity of phytoplankton in Lake Tahoe, California. Goldman (1974) had previously shown that Lake Tahoe is subject to rapid and accelerated eutrophication and the analysis of Goldman's data by Dennis *et al.* shows positive relationships between species richness and diversity of the phytoplankton and water temperature.

Application of diversity indices

The nature and application of diversity indices and the advantages of one index over another have been discussed for decades (see, for example, Hurlbert 1971, Peet 1974, Usher 1983, Wolda 1983, Gadagkar 1989). It is especially useful to consider the advantages and disadvantages of using diversity indices, particularly in monitoring programmes or in relation to measurements of impacts on the natural environment. Diversity indices are simple mathematical expressions and from that point of view there is nothing special let alone mystical about them. One advantage of an index is that a lot of data can be neatly summarized and recorded in one figure or set of figures. An expert witness in a public enquiry may seem to be giving very impressive evidence in the form of indices but indices do need to be put into context and they also need to be qualified. That is, diversity indices tell us nothing about the type of distribution, stage of succession or species composition.

An index score is meaningless unless it can be put into context and therefore we should ask, what is the minimum diversity index and the maximum diversity index which could be expected given certain conditions? In order to put values into perspective, a good way is to calculate hypothetical maximum and hypothetical minimum values. We might also ask, how consistent are the diversity values? In other words what degree of statistical variance is there when the diversity is measured on several occasions? Variance of the diversity indices should be made known or at least some measure of variation should be noted.

Although an index has the advantage of being able to summarize a large amount of data in a simple form, it is perhaps useful to consider qualifying an index by using other data, especially when monitoring changes in biotic communities. The number of species, numbers of individuals, total biomass and biomass at various trophic levels are equally valuable attributes which can be used in monitoring.

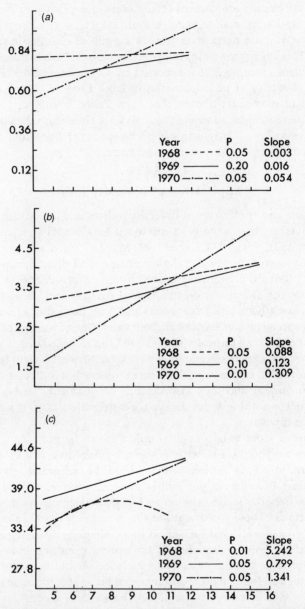

Fig. 6.5. Species richness and diversity of phytoplankton as related to temperature for Lake Tahoe, California: (a) diversity calculated with Simpson's Index; (b) diversity calculated with the Shannon–Wiener Index; (c) species richness or number of species. These graphs were prepared and modified from regression analysis by Dennis et al. (1979) using data from Goldman (1974).

It has long been known that for various taxonomic groups, the number of species increases with increasing sample size or effort, that is there is a species–area relationship (Williamson 1981). This relationship has important implications when interpreting species richness and species diversity. The biological explanations for the species–area relationship and the mathematical descriptions of the relationship have been well researched and reviewed (Spellerberg 1991). For example, Preston (1948) and MacArthur & Wilson (1967) developed models to explain species richness on islands of various sizes as a function of immigration and extinction rates. Williams (1964) developed the habitat diversity hypothesis which assumes that as area increases, so more habitats are included and thus more species are recorded.

The species–area relationship can pose a serious problem in the use of diversity indices, that is we need to know if the sample size has any effect on the index. This can easily be investigated by pooling successive standard samples (i.e. from quadrats and not sweeps) and then calculating the index for each set of data. One example showing a small effect of increased sample size when using the Shannon–Wiener Index is clearly shown in Fig. 6.6. As a guide, a minimum of ten samples would be taken.

While diversity indices have had popular use in ecological monitoring, there are many conceptual, semantic and technical problems (Hurlbert 1971). For example, the use of a species diversity index assumes that all

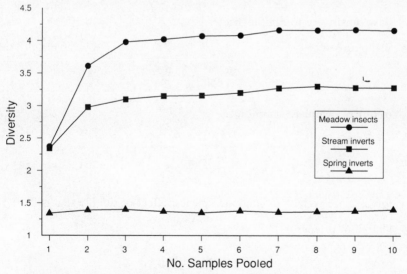

Fig. 6.6. Species diversity (Shannon–Wiener) values obtained by pooling successive samples from one through to ten. Data from Wilhm & Dorris (1968).

species have been sampled. This can rarely be the case and therefore the index will always be biassed towards an under-estimate of the diversity.

Only a few diversity indices have been described here but a very comprehensive and clear account may be found in Magurran (1988). She provides examples of a wide range of diversity indices with many worked examples. The few diversity indices mentioned here may be summarized as follows.

1. Hill's

$$D = \frac{(\sum x)^2}{\sum x^2}$$

(one example of a series which may be difficult to interpret and which has not been widely used)

2. Sequential

$$D = \frac{x}{N}$$

(simple but attention to randomness of sample required; high sensitivity to sample size)

3. Williams's alpha

$$S = \alpha \log_e(1 + N/\alpha)$$

(low sensitivity to sample size)

4. Brillouin's

$$H = \frac{1}{N} \log_{10} \frac{N!}{N_1! N_2! N_3!}$$

(requires complete sampling)

5. Simpson's (Yules)

$$D = \sum_{i=1}^{s} P_i^2$$

(gives more weight to common species; effect of sample size is low)

6. Shannon–Wiener

$$D = -\sum_{i=1}^{s} P_i^2(\log_e P_i)$$

(strongly influenced by species richness but effect of sample size is low)

References

Belsky, A. J. (1985). Long-term vegetation monitoring in the Serengeti National Park, Tanzania. *Journal of Applied Ecology*, **22**, 449–60.

Ben-Eliahu, M. N. & Safriel, U. N. (1982). A comparison between species diversities of polychaetes from tropical and temperate structurally similar intertidal habitats. *Journal of Biogeography*, **9**, 371–90.

Brillouin, L. (1960). *Science and Information Theory*, 2nd ed. New York, Academic Press.

Cairns, J., Albaugh, D. W., Busey, F. & Chanay, M. D. (1968). The sequential comparison index – a simplified method for non-biologists to estimate relative differences in biological diversity in stream pollution studies. *Journal of Water Pollution Control Federation*, **40**, 1607–13.

Dennis, B., Patil, G. P. & Rossi, O. (1979). The sensitivity of ecological diversity indices to the presence of pollutants in aquatic communities. In *Environmental Biomonitoring, Assessment, Prediction and Management – Certain Case Studies and Related Quantitative Issues*, ed. J. Cairns, G. P. Patil & W. E. Waters, pp. 379–413. Maryland, International Cooperative Publishing House.

Dills, G. & Rogers, D. T. (1974). Macroinvertebrate community structure as an indicator of acid mine pollution. *Environmental Pollution*, **6**, 239–62.

Fisher, R. A., Corbet, A. S. & Williams, C. B. (1943). The relation between the number of species and the number of individuals in a random sample of an animal population. *Journal of Animal Ecology*, **12**, 42–58.

Gadagkar, R. (1989). An undesirable property of Hill's diversity index N_2. *Oecologia*, **80**, 140–1.

Goldman, C. R. (1974). *Eutrophication of Lake Tahoe emphasizing water quality*. Report No. EPA-660/3-74-034, *Ecological Research Series*, U.S. Environmental Protection Agency.

Hellawell, J. M. (1977). Change in natural and managed ecosystems: detection, measurement and assessment. *Proceedings of the Royal Society of London* B**197**, 31–57.

Hill, M. O. (1973). Diversity and evenness: a unifying notation and its consequences. *Ecology*, **54**, 427–32.

Hurlbert, S. H. (1971). The non-concept of species diversity: a critique and alternative parameters. *Ecology*, **54**, 427–32.

Jones, A. (1979). *The effects of people trampling on the vegetation of Balmer Lawn*. Southampton University, Unpublished Undergraduate Environmental Sciences Project.

Liddle, M. J. (1975). A selective review of the ecological effects of human trampling on natural ecosystems. *Biological Conservation*, **7**, 17–36.

MacArthur, R. & Wilson, E. O. (1967). *The Theory of Island Biogeography*. New Jersey, Princeton University Press.

Magurran, A. E. (1988). *Ecological diversity and its Measurement*. London, Croom Helm.

Margalef, R. (1951). Diversidad de especies en las comunidades naturales. *PubInes. inst. Biol. apl., Barcelona*, **6**, 59–72.

Menhinick, E. F. (1964). A comparison of some species–individuals diversity indices applied to samples of field insects. *Ecology*, **45**, 859–61.

Peet, R. K. (1974). The measurement of species diversity. *Annual Review of Ecology and Systematics*, **5**, 285–307.

Pielou, E. C. (1966). The measurement of diversity in different types of biological collections. *Journal of Theoretical Ecology*, **13**, 131–44.

Preston, F. W. (1948). The commonness and rarity of species. *Ecology*, **29**, 254–83.

Simpson, E. H. (1949). Measurement of diversity. *Nature*, **163**, 688.

Southwood, T. R. E. (1966). *Ecological Methods with Particular Reference to the Study of Insect Populations*. London, Chapman & Hall.

Spellerberg, I. F. (1991). A biogeographical basis of conservation. In *The Scientific Management of Temperate Communities for Conservation*, pp. 293–322, ed. I. F. Spellerberg, F. B. Goldsmith & M. G. Morris. Oxford, Blackwells.

Usher, M. B. (1983). Species diversity: a comment on a paper by W. B. Yapp. *Field Studies*, **5**, 825–32.

Wilhm, J. L. & Dorris, T. C. (1968). Biological parameters for water quality criteria. *BioScience*, **18**, 477–481.

Williams, C. B. (1964). *Patterns in the Balance of Nature*. New York, Academic Press.

Williams, N. V. & Dussart, G. B. J. (1976). A field course survey of three English river systems. *Journal of Biological Education*, **10**, 4–14.

Williamson, M. (1981). *Island Populations*. Oxford, Oxford University Press.

Wolda, H. (1983). Diversity, diversity indices and tropical cockroaches. *Oecologia*, **58**, 290–8.

7
Similarity

Introduction

ECOLOGISTS and environmental managers have been looking at ways of quantifying the extent of similarity and dissimilarity between communities or biotic samples for more than 80 years. During that time, a number of coefficients of similarity, dissimilarity and association have been devised (see Clifford & Stephenson 1975 for a review). Similarity indices have had popular use in field ecology, particularly as a basis for quantifying similarities between communities and in the investigation of discontinuities between different sampling methods employed at any one particular site. These indices can also be used in comparing communities which have been subjected to pollution but the assumption has to be made that the communities being compared will have received the same types of pollutants. Comparison of samples from the same community taken at various intervals of time is one basis for the application of similarity indices in monitoring. Also, analysis of community structure, using similarity indices and cluster analysis provides a basis for monitoring changes in a community.

Community similarity indices

Similarity indices have also been used as a basis for cluster analysis, showing affinities between samples and thus providing a baseline for monitoring programmes. Thus, similarity indices can be used for the analysis of data from various communities but can also be used to quantify differences between successive samples from one community. For convenience, the term 'sample' is used but this could also mean 'community'.

Simple but effective, binary community similarity indices, and those which are based only on the number of species whether common or rare, include Sorensen's Index (Sorensen 1948) and Jaccard's Index (Jaccard 1902). The latter was originally known as the Coeffieient of Floral Community and is probably the earliest index of similarity. These indices are calcuated as follows:

$$C = \frac{2w}{A+B} \cdot 100 \quad \text{(Sorensen)}$$

$$J = \frac{w}{A+B-w} \cdot 100 \quad \text{(Jaccard)}$$

where C or J is the index of similarity, w is the number of species common to both samples (or community) and A is the number of species in sample one and B is the number of species in sample two. Here the scores have been multiplied by 100 to give a percentage scale. A value of zero per cent indicates complete dissimilarity whereas a value of 100 per cent indicates maximum similarity between samples (or communities). Both these indices increase with increasing sample size (Mountford 1962). As they both give equal weight to all species they place too much significance on the rare species in the samples.

The abundance of the different species can be considered using a modification of the Sorensen Index:

$$C_n = \frac{2jN}{(aN+bN)}$$

where aN is the total number of individuals in sample A, bN is the total number of individuals in sample B and jN is the sum of the lower of the two abundances recorded for species found in both samples. That is if 10 individuals of one species were in community A, but in sample B there were 20 individuals of the same species, then the value of 10 would be included in the sum and so on.

An index of community similarity based on percentages is the Percentage Community Similarity Index (Brock 1977) and this is expressed by:

$$P_{SC} = 100 - 0.5\sum(a-b)$$

where P_{SC} is the index of community similarity, a and b are for given samples, percentages of the total sample. The absolute value of the differences is summed overall. For example looking at the data in Table 7.1, the P_{SC} for samples A and C would be calculated as follows:

A (%)	C (%)	Difference between % of a and b
4 (1.6)	2 (0.9)	0.7
60 (23.8)	57 (26.0)	2.2
1 (0.4)	0 –	0.4
65 (25.8)	83 (37.8)	12.1
25 (9.9)	44 (20.1)	10.2
etc.		Total 55.5

$$P_{SC} = 100 - 0.5 \ (55.5)$$

$$= 72.25$$

Table 7.1. *Benthic macro-invertebrate samples taken at monthly intervals from one stream*

Species	Samples and abundance of species					
	A	A¹	B	C	D	E
Oligochaeta						
Haplotaxius sp.	4	4	3	2	2	0
Lumbriculus variegatus	60	64	54	57	51	0
Stylodrilus sp.	1	2	0	0	6	0
Tubifex sp.	65	66	81	83	196	108
Chaetogaster sp.	25	24	40	44	31	0
Crustacea						
Ascellus aquaticus	3	3	2	3	1	0
Gammarus pulex	19	20	2	4	3	1
Insecta, Diptera						
Chironomous sp.	5	3	20	11	30	18
Spaniotoma sp.	21	18	4	5	4	0
Tanypus sp.	46	45	5	10	10	0
Insecta, Ephemeroptera						
Ephemera sp.	3	3	1	0	0	0
Species richness	11	11	10	9	10	3
Number of individuals	252	252	212	219	334	127
Community Similarity Indices						
Sorensen		1.0	0.95	0.9	0.95	0.43
P_{SC}		97.2	69.5	72.3	59.4	60.9
Pinkham & Pearson		0.8	0.41	0.44	0.33	0.08
Morista		1.0	0.8	0.9	0.7	0.48

Samples A¹ and A were taken at about the same time to provide data for a baseline similarity index. For calculation of similarity indices, A¹, B, C, D and E are compared with A. Small quantities of industrial effluents entered the stream after samples A and A¹ had been taken and a large pollution incident occurred in the interval between samples D and E.

During the late 1970s, an extensive programme of research on Water Boatmen (*Corixidae*) led Savage to devise an index of similarity based on percentages (proportions) because he intended to reflect relative numbers of individuals of different species in different habitats and not actual numbers (see, for example, Savage & Pratt 1976 and Savage 1982). In connection with that research on water boatmen, Savage looked at the classification of lakes and used a form of a similarity index as follows:

$$I = \frac{[(a_1 \times b_1) + (a_2 \times b_2) \dots (a_n \times b_n)]^2}{(a_1^2 + a_2^2 \dots a_n^2) \cdot (b_1^2 + b_2^2 + \dots b_n^2)}$$

where I is the Index of Similarity and a_1 is the percentage of individuals of a given species at a particular site and b_1 the percentage of individuals of the same species at the other site ('sites' could be replaced by 'samples' in monitoring or surveillance); a_2 and b_2 are the respective percentages for the second species and so on up to any number of species. For example and with reference to two lakes (Hanmer Mere, Oak Mere) and six species of the Corixid *Sigara* the index would be calculated as follows (Savage 1982):

	scotti	distincta	concinna	praeusta	dorsalis	falleni
Hanmer Mere	0% (a_1)	26.7% (a_2)	1.3% (a_3)	17.3% (a_4)	18.7% (a_5)	36.0% (a_6)
Oak Mere	1.5% (b_1)	2.2% (b_2)	0% (b_3)	9.7% (b_4)	85.1% (b_5)	1.5% (b_6)

$$I = \frac{[(0 \times 1.5) + (26.7 \times 2.2) \dots]^2}{(0^2 + 26.7^2 \dots) \cdot (1.5^2 + 2.2^2 \dots)}$$

$$= 0.1$$

If the same species are present in identical percentages at two sites then the index of similarity is 1.0; if there are no species in common then the index would be 0.0. Further comments about the application of Savage's work are made in Chapter 12.

This index of similarity can be used with numbers of individuals rather than percentages. For example, Morris (1969) used the above index in the form:

$$I = \frac{[\sum(m_i \times n_i)]^2}{\sum m_i^2 \times \sum n_i^2}$$

where m_i is the number of specimens of species i recorded from the first site or sample and n_i is a similar term for species from the second site or second sample.

The Pinkham and Pearson Index of Community Similarity (Pinkham & Pearson 1976) is based on actual and not relative abundance of the species and is expressed by the following:

$$P = \frac{1}{k} \sum \left[\frac{\min(X_{ia} \cdot X_{ib})}{\max (X_{ia} \cdot X_{ib})} \right]$$

where P is the index of community similarity, X_{ia} and X_{ib} are the abundance of species i for the respective samples (the smaller number being divided by the larger for each species), k is the total number of comparisons or different taxa in the two samples.

The Morista Index of Community Similarity (Morista 1959), although more complex in its calculation, has advantages with respect to sample size and is expressed as follows:

$$C = \frac{2 \sum n_{1i} n_{2i}}{(\lambda_1 + \lambda_2) N_1 N_2}$$

where

$$\lambda_1 = \frac{\sum n_{ji} (n_{ji} - 1)}{N_j (N_j - 1)}$$

and C is the index of community similarity, N is the number of individuals in sample j, n_{ji} is the number of individuals of species i in sample j.

A 'sub-sample' from data on stream macro-invertebrates provides us with some data with which to demonstrate the use of Sorensen's, Pinkham and Pearson's, and the Morista community similarity indices (see Table 7.1). Note that five samples were collected from the same site at intervals of one month and an additional one sample (A^1) was taken within a few days of the first sample (A) to act as a baseline in the interpretation of the results. After the first two samples were taken, small quantities of industrial effluent entered the stream and then in the interval between samples D and E being collected, larger amounts of effluent entered the stream. The number of species, the number of individuals and community similarity indices are shown.

The Sorensen Index of Community Similarity is easily calculated from the number of species only. Looking at the information in Table 7.1, it is perhaps not surprising that samples B, C and D are very similar to A but that E is clearly dissimilar. The same trend is obtained for the Percentage Community Similarity Index.

In the case of the Pinkham and Pearson Index of Community Similarity, the method of calculation is as follows for samples A and B:

$$P = \frac{1}{11} \sum \frac{3}{4} + \frac{54}{60} + \frac{0}{1} + \frac{65}{81} \cdots$$

$$= 0.41$$

The Morista Index of Community Similarity for samples A and B (Table 7.1) is calculated as follows:

$$\lambda_i = \frac{(4 \times 3) + (60 \times 59) + (1 \times 0) \ldots}{252 \times 251}$$

$$C = \frac{2 \sum (4 \times 3) + (60 \times 54) + (1 \times 0)}{(0.429) \times 252 \times 212}$$

Application of community similarity indices

When interpreting indices of community similarity, it is important to have a baseline index and information on the sample size. That is, in addition to calculating the indices for samples taken at various intervals in time, a value indicating maximum similarity should be calculated from two samples taken at one station at about the same time. This provides a baseline indes with which to compare indices calculated at later intervals. For example and with reference to the benthic macro-invertebrates (Table 7.1). A and A[1] are two such samples which were taken as a basis for calculating a baseline similarity index.

Sample size does affect similarity indices and research on this aspect has been undertaken in some detail by Wolda (1981) who concluded that the Morista Index was least affected by sample size and species richness. In another detailed analysis Smith (1986) favoured the use of the Sorensen Index as a good binary index of community similarity and various forms of the Morista Index as a good quantitative index.

The Pinkham and Pearson Index is particularly sensitive to rare species and has been criticized as not being sensitive enough to variation in the dominant species (Wolda 1981). By way of comparison, the P_{SC} index has a greater response to variation in dominant forms and the relationship between dominant and semi-dominant forms. As with the species diversity indices, it would seem not only useful to make a careful selection of the index but also to employ more than one index in case any particular one method gives misleading results.

Ranking, classification and ordination

The ranks of the proportional abundance of different taxa in two samples can provide a basis for comparison, using non-parametric tests such as Spearman's or Kendall's rank correlation coefficients (see Huhta 1979 for application of Kendall's rank correlation test). Although the former test is described here for completeness, comprehensive statistical books should be consulted, e.g. Sokal & Rohlf (1981). Absolute measurements of species

Table 7.2. *Comparison of the relative abundance of plant species in an unmown grass sward and a sward mown once a year in August*

Species	Mean percentage cover Unmown	Mean percentage cover August mown	Rank Unmown	Rank August mown	d
Holcus lanatus	32.4	30.8	1	1	0
Dactylis glomerata	15.0	13.7	2	3	1
Festuca rubra	9.1	13.0	3	4	1
Agrostis sp.	8.3	7.6	4	5	1
Plantago lanceolata	8.2	1.5	5	9	4
Arrhenatherum elatius	5.9	0.3	6	12	6
Centaurea nigra	4.6	15.0	7	2	5
Malva moschata	2.8	0.0	8	18	10
Heracleum sphondylium	0.6	1.2	9.5	10	0.5
Rumex acetosa	0.6	0.0	9.5	18	8.5
Hypochaeris radicata	0.5	0.3	11	12	1
Phleum pratense	0.4	0.2	13.5	14	0.5
Anthoxanthum odoratum	0.4	3.2	13.5	7	6.5
Ranunculus acris	0.4	0.0	13.5	18	4.5
Trifolium pratense	0.4	0.3	13.5	12	1.5
Vicia sativa	0.3	0.1	16	15.5	0.5
Ranunculus bulbosus	0.1	0.1	17	15.5	1.5
Lotus corniculatus		5.0	18.5	6	12.5
Achillea millefolium		1.8	18.5	8	10.5
Litter	9.6	5.3			
Bare ground	0.3	0.6			

Data from an Environmental Sciences Student Project (Bridget Smith), Southampton University.

abundance are not required, since it is the relative importance of each species that is used in the calculation of the rank correlation. This is an advantage in monitoring programmes when large samples or a large sampling effort is required.

Spearman's rank correlation is calculated as follows:

$$r_s = 1 - \frac{6 \sum d^2}{n^3 - n}$$

where r_s (-1 to $+1$) is the correlation coefficient, d is the difference in the magnitude of the rank of a species or group of taxa in the two samples and n is the total number of species or groups of taxa in the paired comparison. Ties of rank are given mean values and where a species occurs in the first group but not in the second, it is put last in the rank of the second list. Testing whether or not there is a statistical difference between samples is possible but depends on the number of species involved (see any basic statistical book).

Using data in Table 7.2, Spearman's rank correlation would be calculated as follows:

$$r_s = 1 - 6 \times 587.5/6840$$

$$= 1 - 0.515$$

$$= 0.48$$

Classification of data by way of cluster analysis (Sneath & Sokal 1973, p. 192 *et seq.*) provides a visual or graphic method for examining and interpreting levels of affinities between groups. In other words, on a collection of samples we may wish to determine which are most closely related on the basis of similarity indices. This could be useful for establishing baseline data on affinities between communities and later, similar analysis may indicate the extent to which perturbations have affected the affinities. There are many kinds of cluster analysis and one good example of the use of this method of analysis is the work of Savage (1982) who followed a method of analysis previously used by Mountford (1962). Indices of similarity were used to construct a dendrogram to show the affinities between communities of Water Boatmen (Corixidae) in a series of lakes. The details of the calculation are given in Appendix I.

Cluster analysis using the data from the Jaccard Index of community similarity has had popular use in research on pollution of freshwater communities. For example, in an examination of the effects of pollutants on

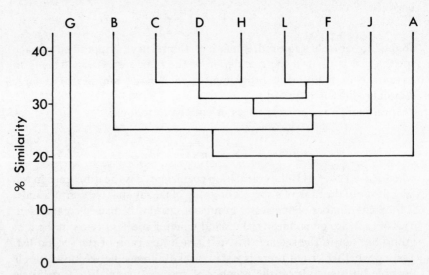

Fig. 7.1. A classification (dendrogram) showing station similarity based on average monthly Jaccard similarity indices. Redrawn from Scullion & Edwards (1980).

the macro-invertebrate fauna in a small river, Scullion & Edwards (1980) assessed various methods as a basis for biological surveillance. The Jaccard Index was used to quantify the extent of similarity between sampling stations. Arithmetic means of monthly Jaccard indices were then calculated for each pair of sampling stations and then clustered using the average linkage method (Sneath & Sokal 1973). The results of this analysis can then be expressed in the form of a dendrogram as shown in Fig. 7.1. The conventional dendrogram is in the form of a rectangular grid with the level of affinity on the ordinate and the groups on the abscissa. In Fig. 7.1 it can be seen that the stations C, D and H form one level of affinity. These were the unpolluted sampling stations while sampling stations E and F were influenced by coal siltation and G and B (which the analysis has not grouped together) by acid pollution.

Despite the wide use of cluster analysis as a method of examining similarity measures, there has been some criticism of the apparent and not real objectivity of cluster analysis. In an assessment of 11 similarity measures used to assess the impacts of clam digging on the infauna of an intertidal mudflat, Hruby (1987) concluded that too many subjective choices are made in selecting algorithms and methods of interpretation. To reduce the number of subjective choices made, Hruby used ANOVA of the similarity matrix as an alternative to clustering.

Classification of communities has also been undertaken using Two-Way Indicator Species Analysis (TWINSPAN). One interesting example which

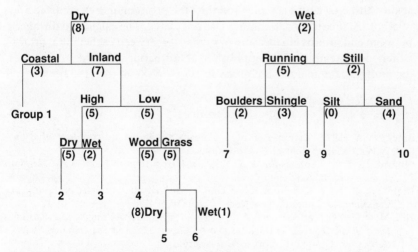

Fig. 7.2. Classification (by TWINSPAN) of 248 Carabid localities in north-east England. The numbers in brackets refer to the number of indicator species at each division. Redrawn from Luff *et al.* 1989.

uses this method is the classification of Carabid Beetles from 248 localities in north-east England undertaken by Luff, Eyre & Rushton (1989). TWINSPAN is a FORTRAN programme for arranging multivariate data in ordered two-way tables (Hill 1979) and it produces a hierarchical classification by repeatedly dividing data into groups. The classification of Carabid Beetles by Luff *et al.* (1989) produced a hierarchical classification by dividing the data into two groups, followed by repeated division of each group and this was continued until each further division produced two end groups which were not ecologically distinct. The analysis resulted in the identification of ten ecologically meaningful habitat groups (Fig. 7.2) which could usefully serve as a baseline data set for a monitoring programme.

Ordination is a multivariate technique of analysis particularly suitable for large data sets and is now commonly used in ecological research and analysis but less so in monitoring studies. Nevertheless the technique can usefully be noted here, along with ranking and classification. The multidimensional arrangement of both sites and species in the same multidimensional space can be used as a basis for describing site characteristics of importance to community structure or it can be used to describe preferred environments of those species present. For example, Luff *et al.* (1989) used the FORTRAN programme known as DECORANA (Detrended Correspondence Analysis) to ordinate the species of carabids from the 248 localities.

Using this analysis, Luff *et al.* plotted the centroids of each habitat group against three DECORANA axes and then the position of each centroid and relative distance from one another was calculated. The calculated distances between end groups is a measure of similarity between each habitat group. Other good examples of the application of ordination and other techniques for multivariate analysis are given in Digby & Kempton (1987).

References

Brock, D. A. (1977). Comparison of community similarity indices. *Journal of the Water Pollution Control Federation*, **49**, 2488–94.

Clifford, H. T. & Stephenson, W. (1975). *An Introduction to Numerical Classification*. London, Academic Press.

Digby, P. G. N. & Kempton, R. A. (1987). *Multivariate Analysis of Ecological Communities*. London, New York, Chapman & Hall.

Hill, M. O. (1979). TWINSPAN – *A* FORTRAN *Program for Arranging Multivariate Data in an Ordered Two-way Table by Classification of the Individuals and Attributes*. New York, Cornell University.

Hruby, T. (1987). Using similarity measures in benthic impact assessments. *Environmental Monit`ring and Assessment*, **8**, 163–80.

Huhta, V. (1979). Evaluation of different similarity indices as measures of succession in arthropod communities of the forest floor after clear-cutting. *Oecologia*, **41**, 11–23.

Jaccard, P. (1902). Lois de Distribution Florale dans la Zone Alpine. *Soc. Vaud. Sci. Natl. Bull.*, **38**, 69.

Luff, M. L., Eyre, M. D. & Rushton, S. P. (1989). Classification and ordination of habitats of ground beetles (Coleoptera, Carabidae) in north-east England. *Journal of Biogeography*, **16**, 121–30.

Morista, M. (1959). Measuring of interspecific association and similarity between communities. *Memoirs Faculty of Science Kyushu University, Series E, Bio.*, **3**, 65–80.

Morris, M. G. (1969). Associations of aquatic Heteroptera at Woodwalton Fen, Huntingdonshire, and their use in characterising aquatic biotopes. *Journal of Applied Ecology*, **6**, 359–73.

Mountford, M. D. (1962). An index of similarity and its application to classificatory problems. In *Progress in Soil Zoology*, ed. P. W. Murphy, pp. 43–50. London, Butterworths.

Pinkham, C. F. A. & Pearson, J. G. (1976). Applications of a new coefficient of similarity to pollution surveys. *Journal of Water Pollution Control Federation*, **48**, 717–23.

Savage, A. A. (1982). Use of water boatmen (Corixidae) in the classification of lakes. *Biological Conservation*, **23**, 55–70.

Savage, A. A. & Pratt, M. M. (1976). Corixidae (water boatmen) of the Northwest Midland meres. *Field Studies*, **4**, 465–76.

Scullion, J. & Edwards, R. W. (1980). The effects of coal industry pollutants on the macro-invertebrate fauna of a small river in the South Wales coalfield. *Freshwater Biology*, **10**, 141–62.

Smith, B. (1986). *Evaluation of Different Similarity Indices Applied to Data from the Rothamsted Insect Survey.* Unpublished M.Sc. Thesis, University of York.

Sneath, P. H. A. & Sokal, R. R. (1973). *Numerical Taxonomy. The Principles and Practice of Numerical Classification.* San Francisco, Freeman & Co.

Sokal, R. R. & Rohlf, F. J. (1981). *Biometry. The Principles and Practice of Statistics in Biological Research*, 2nd ed., San Francisco, Freeman & Co.

Sorensen, T. A. (1948). A method of establishing groups of equal amplitude in plant sociology based on similarity of species content, and its application to analyses of the vegetation on Danish commons. *K. dan. Vidensk, Selsk. Biol. Skr.*, **5**, 1–34.

Wolda, H. (1981). Similarity indices, sample size and diversity. *Oecologia*, **50**, 296–302.

8

Environmental and biotic indices

Introduction

WE HAVE PREVIOUSLY REFERRED TO two kinds of quantitative indices, that is diversity and similarity indices (Chapters 6 and 7) but there are many kinds of other biological indices and many kinds of environmental indices. Indeed, almost every kind of discipline has made use of various indices and therefore we should understand what an index is. An index, in the sense used here, is a number or a quantity compared to an arbitrary standard. In economics, for example, a price index is a weighted average of prices of consumer goods and services produced in the economy measured over time and one example is the Retail Price Index.

Environmental indices which have been used to assess and monitor environmental quality include not only those used in relation to pollution but also those used to assess and monitor landscapes and aesthetics. There are many ways of assessing landscape quality and a number of schemes incorporating scoring systems have been developed (Ribe 1986, Tips & Savasdisara 1986). These scores of landscape quality can be used in the sense of an index and thus be used to monitor changes. Other environmental indices include those for noise, radioactivity, air quality and water quality (see Thomas 1972 for a review of environmental indices).

Environmental indices

A whole range of environmental indices can form the basis of assessing environmental quality. A widely accepted method of normalization has proved popular in these circumstances, particularly where the quality of wildlife, air, noise, dust, recreation and other environmental variables are being assessed. That is, interpretation of the environmental variable can be achieved by converting measured variables onto normalized numbers or

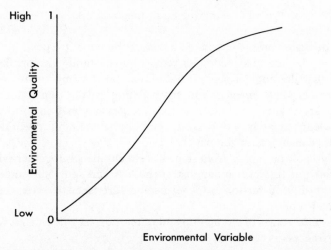

Fig. 8.1. Normalization or conversion of a hypothetical environmental variable (such as SO_2 levels or species richness) to levels of environmental quality (scored from 0 to 1).

an index by means of a uniform scale. A hypothetical example is shown in Fig. 8.1 and a further example is shown in Fig. 15.2. Kreisel (1984) proposed this method, that is the use of empirically derived indices, for describing spatially and temporally varying environmental quality of a metropolitan area (the City of Dortmund). Environmental variables were selected from three categories (air pollution, noise pollution and recreational facilities as one example of man-made environments) and each variable was then normalized on a 0 to 1 scale. To derive composite indices, the normalized variables or indicators of environmental quality were weighted, using a method based on the Delphi technique and then aggregated. The results were then converted to isopleths plotted on to a map of Dortmund.

The Delphi technique (Linstone & Turoff 1975, Sackman 1975) as used by Kreisel for making numerical comparisons and weighting is worth noting in some more detail. The Delphi technique was developed in the 1960s by Helmer at the Rand Corporation as one of the many techniques used in decision analysis, probability estimates and long-range forecasts. The technique is sometimes used as a means for promoting a consensus of the views expressed by a panel of experts or a committee. Basically, Delphi aims to encourage independent and unbiassed assessments from each individual.

In brief, the technique uses a questionnaire which is circulated amongst a panel of experts who are not aware of the identity of fellow members of the panel. There are several stages as follows.

1. A questionnaire addressing the environmental issues is designed by a small team.
2. The questionnaire is circulated amongst the panel by post.
3. The replies are analysed then redistributed stating the median and interquartile range of replies. Individual panel members are asked to reconsider their answers and those falling outside the interquartile range are invited to state their reasons. Reasons may include lack of knowledge or by way of contrast some specialist knowledge unknown to other members of the panel.
4. Analysis of the replies from stage 2 are recirculated, together with the reasons put forward in support of the 'extreme' positions and in the light of this information the panel members are asked to reconsider their original replies.
5. Additional stages of circulation and reiteration are undertaken when it is felt necessary to do so.

The success of this and other kinds of techniques for decision analysis is very much dependent on the selection of the panel members and the design of the questionnaire. Although Delphi has been widely used for environmental assessment and for monitoring programmes (Linstone & Turoff 1975, Richey, Mar & Horner 1985) this technique, like the many others, does have its critics (Sackman 1975).

Wildlife indices

In 1971, the MITRE Corporation in the USA prepared a report on environmental indices as a background report for the Council on Environmental Quality. That report by the MITRE Corporation considered a total of 112 indices (together with detailed calculations and equations) ranging from indices for odours in the home to fish kills and also wildlife indices (Inhaber 1976). The MITRE report suggested two types of wildlife indices, one for endangered species (those species which are approaching extinction) and one for 'troubled' species (species which have been greatly reduced in numbers but some populations still thrive). The calculation of these wildlife indices takes into account a subjective weighting of a species' position in a food chain with endangered carnivores scoring 5 and rodents 1. This is because, it was argued, changes in the groups higher in the food chain indicate effects that have occurred lower down.

Many other wildlife indices have been developed and one recent example is the Threat Number (Perring & Farrell 1983) which was devised to assess the conservation status of endangered flora in Britain (see p. 53). Although not specifically devised for monitoring the status of plants, threat numbers could be used with other data in monitoring the extent of degradation in natural and semi-natural areas. Wildlife indices and species evaluations in general (Spellerberg 1981) have an important role in assessing the status of a species and focusing attention on the conservation needs of those species.

Biotic indices

Monitoring programmes can generate very large amounts of data, so large that there may be considerable demands on laboratory time for sorting of the material, let alone analysis of the data. Biological monitoring programmes are not uncommonly part of an interdisciplinary management programme and a further problem may arise in that it is sometimes difficult to summarize the results in a manner which is understandable to non-biologists. Diversity indices and similarity indices can require complicated calculations and the index may seem distant from the biological attributes of the community being examined. Summary and communication problems can be overcome by using one of several biotic indices which are based on biological attributes such as sensitivity to certain kinds of pollution.

As with diversity and similarity indices, there are many biotic indices (see, for example, Hellawell 1978, 1986). The examples selected and outlined in this section are biassed towards the monitoring of freshwater and applications of these indices are discussed later in Chapter 12.

The Neville Williams Index

There have been many biological studies of polluted rivers and lakes and therefore the effects of organic pollution in particular have been well documented for many years (see, for example, Hynes 1960, Abel 1989). Those studies have led to the development of a number of indices based largely on the effects of organic pollution on species composition. For example, the Neville Williams Index is based on a subjective classification of taxonomic groups into those which are tolerant, intolerant and indifferent to pollution. It was an index which received wide use in the earlier attempts to monitor water quality. For aquatic communities, tolerant groups would include leeches, some molluscs and some Oligochaeta. Intolerant groups would include Ephemeroptera, Trichoptera, and Plecoptera. A simple, similar dichotomy could be assembled for other groups such as lichens (in relation to airborne pollution), grasses (in relation to trampling) and birds (in relation to vegetation structure and physical disturbance). In New Zealand, for example, Dacre & Scott (1973) used the numerical ratio of Ephemeroptera to Oligochaeta as a basis for an index and despite some criticism as to the simplicity of such an index, such a simple approach has received some support in connection with monitoring rivers in New Zealand (Scott 1989).

The Neville Williams Index is calculated as follows:

$$I = \text{per cent tolerant organisms/per cent intolerant}$$

where I is the index. Values greater than one indicate polluted or disturbed conditions while values less than one would indicate conditions free of pollution or disturbance.

These and other indices can usefully be expressed in the manner suggested by Williams & Dussart (1976), that is, in the form of clock diagrams where the extent of shading indicates the degree of disturbance or pollution (Fig. 8.2).

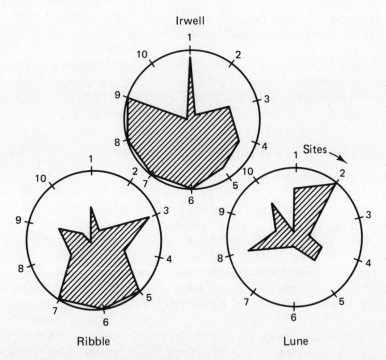

Fig. 8.2. Clock diagrams comparing the Neville Williams biotic index for water quality in three rivers in England (The Irwell, Ribble and Lune). The smaller the central shaded area, the less polluted the water. Redrawn from Williams & Dussart (1976).

Table 8.1. *Allocation of biological scores (the BMWP Scores)*

Families	Score
Siphlonuridae Heptageniidae Leptophlebidae Ephemerellidae Potamanthidae Ephemeridae Taeniopterygidae Leucridae Capniidae Periodidae Periidae Chloroperlidae Aphelocheiridae Phryganeidae Molannidae Beraeidae Odontoceridae Leptoceridae Gooeridae Lepidostomatidae Brachycentridae Sericostomatidae	10
Astacidae Lestidae Agriidse Gomphidae Cordulegasteridae Aeshnidae Corduliidae Libeliuidae Psychomyiidae Philopotamidae	8
Caenidae Nemouridae Rhyacophilidae Polycentropodidae Limnephilidae	7
Neritidae Viviparidae Ancylidae Hydroptilidae Unionidae Corophiidae Gammaridae Platycnemididae Coenagriidae	6
Mesovelidae Hydrometridae Gerridae Nepidae Naucoridae Notonectidae Pleidae Corixidae Haliplidae Hygrobidae Dytiscidae Gyrinidae Hydrophilidae Clambidae Helodidae Dryopidea Elminthidae Chrysomelidae Curculionidae Hydropsychidae Tipulidae Simuliidae Planariidae Dendrocoelidae	5
Baetidae Sialidae Piscicolidae	4
Valvatidae Hydrobiidae Lymnaeidae Physidae Planorbidae Sphaeriidae Glossiphoniidae Hirudidae Erpobdellidae Asellidae	3
Chironomidae	2
Oligochaeta (whole class)	1

From National Water Council (1981) *River Quality: The 1980 Survey and Future Outlook.*

The Biological Monitoring Working Party Score

A straightforward but very effective index is the Biological Monitoring Working Party Score (BMWP). Developed and refined over several years, this index is calculated on the basis of selected invertebrate families incorporated into a system which can then be used to assess the biological condition of a river (National Water Council 1981). The BMWP Score system (Table 8.1) has three simple steps; firstly all macro-invertebrate families present at the site are listed, secondly scores are allocated to those families, and thirdly the scores for all the families are summed to give a cumulative site score. Simple and easy to calculate, the BMWP Score system was developed for use in national river pollution surveys but was also developed in response to economic and logistic constraints.

The Saprobian approach

Biological activity in water has provided an alternative basis for a biotic index for water quality. Developed in Europe, the Saprobian system or Saprobian index was received with wide approval but the index does have its critics. Saprobes are organisms of decomposition and decay and the Saprobian system is based on a series of such groups, each of which is associated with different stages of oxidation in organically enriched water. The distribution of aquatic organisms is determined by many factors but in particular by the level of organic matter in the water and by the level of dissolved oxygen. The classification expresses the degree of independence (of the groups) on decomposing organic nutrients. This saprobity is the state of the water quality with respect to the content of decaying organic material but probably reflects oxygen tolerances as well, especially in the β-mesosaprobic zone (see Fig. 8.3). The saprobic zones and the saprobic index are calculated by first allocating groups of organisms to each of the zones. The index can be calculated by one of several equations (Hellawell 1978) and one example is Pantle and Buck's Index:

$$S = \frac{\sum(h_i S_i)}{\sum h}$$

where S_i is the individual saprobic index for each species and h is the relative abundance according to a scale of estimation (1 rare, 3 frequent and 5 abundant). In the calculation, each species is given a number according to the group to which it belongs; lowest value for a xenosaprobic species then the next highest value for an oligosaprobic species and so on. An example of the calculation for Pantle and Buck's Index is shown in Table 8.2 in

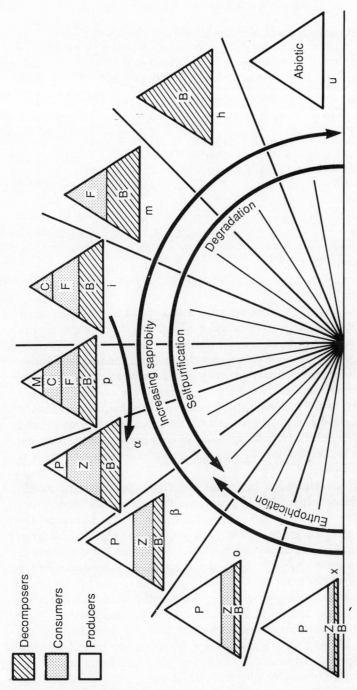

Fig. 8.3. Structure of a saprobic community (x=xenosaprobity, o=oligosaprobity, β=beta-mesosaprobity, α=alpha-mesosaprobity, p=polysaprobity, i=isosaprobity, m=metasaprobity, h=hypersaprobity, u=ultrasaprobity, B=bacteria, F=colourless flagellates, C=ciliates, M=mixotrophic algae and flagellates, Z=zooplankton and other consumers, P=phytoplankton and other producers. Redrawn from Sladecek (1979) with permission of John Wiley & Sons Ltd.

Table 8.2. *Calculation of Pantle and Buck's Index*

Species	Saprobic rating	S_i	Abundance (No.)	h	$h \cdot S_i$
Tubifex tubifex	P	4	2	1	4
Dendrocoelum lacteum	β	2	5	3	6
Asellus aquaticus	α	3	10	3	9
Gammarus pulex	o/β	1 (or 2)	30	5	5
Baetis rhodani	o/β	1 (or 2)	20	5	5
Erpobdella octoculata	α	3	2	1	3
Limnaea stagnalis	β	2	10	3	6
				$\sum 21$	$\sum 38$

$$\text{Index} = \frac{\sum(h_3 S_i)}{\sum h} = \frac{38}{21} = 1.81$$

From Hellawell (1978).

which oligosaprobic organisms are rated 1 and polysaprobic organisms 4. Pantle and Buck's Index gives a value of 1.0 to 1.5 for slight organic pollution and 3.5 to 4.0 for heavy pollution.

In the oligosaprobic zone (clean unpolluted water found near springs) characteristic organisms include Mayfly larvae, mosses such as *Fontinalis antipyretica* and some flat worms such *Planaria gonocephala* (James & Evison 1979). In the α-mesosaprobic zone (water some distance from a pollution source that is poorly oxygenated and contains some nitrites and ammonium salts) characteristic organisms include exploiter-type indicator species such as chironimids, *Tubifex* and larvae of some insects such as *Sialis lutaria* (Alder Fly).

The saprobic method has been criticized for various reasons, mainly because the classification of organisms is not absolute and some organisms may be in one or more zones. Cairns (1979) makes the point that the products of pollution are many and both physical and chemical. A given organism might be intolerant to some kinds of pollution but not others or may be less sensitive to some forms of pollution. This casts doubt on the usefulness of all the biotic indices mentioned here.

In pollution monitoring programmes, biotic indices cannot usefully be interpreted without information on chemical changes in the water. These indices do, however, provide an important, first step in monitoring the effects of pollution. All the indices are based on the same ecological principles (species richness and the extent to which an organism can tolerate exposure to various physical and chemical factors) and whereas the saprobic system can be applied to all freshwater systems and all organisms, the other biotic indices are restricted in their use to running water and aquatic macro-invertebrates. There has been much debate about the uses of the so-called 'British Biotic Indices' and the saprobic indices; in particular, some scientists feel that the biotic indices have limited use because of the use of running water only. For example, Sladecek (1973) who has given strong support to the saprobic system, described the limitations of the biotic indices and gave a fairly convincing demonstration that there is no substantial difference amongst the procedures and results obtained from either the biotic indices or the Saprobian system.

The Trent Biotic Index

In 1964, Woodiwiss described a biotic index which is easy to calculate and simple to understand. This biotic index, which has its origins in the Saprobian system is based on the concept of aquatic indicator species and is weighted by a number of certain defined groups in relation to their sensitivity to organic pollution (Table 8.3). Qualitative samples are taken with a hand net and where possible shallow riffles are used as sampling points. A sampling time of ten minutes was suggested by Woodiwiss (1964).

In the example provided here, we can see how this biotic index (now known as the Trent Biotic Index, because of its introduction by the former Trent River Board in the late 1950s) is used. The lowest score of 0 indicates gross organic pollution and the top score of 15 indicates completely unpolluted conditions. The number of 'groups' (defined in Table 8.3) indicates which column to use. The row to be used is indicated by the presence or absence of certain indicator species. For example, in a sample there may be between 11 and 15 taxonomic groups present and we then look down the appropriate column. In our hypothetical example, there are no Plecopterans but there is one species of Ephemeroptera. The biotic index would therefore be 7. The number 7 by itself is really meaningless because it is possible to arrive at the value of 7 in several different ways; clearly the index value has to be qualified by reference to the various indicator groups.

Overall this index is derived from two effects, the reduction in community richness and the progressive loss of certain detector indicator groups (chosen on a subjective basis) and weighted by the number of defined 'groups'. The index does not require a standardized sampling procedure which overcomes some of the sampling problems outlined in Chapter 4. In its original form, this biotic index had a score ranging from 0 to 10 but in the expanded form (Table 8.3) the scores range from 0 to 15

Table 8.3. *The Trent Biotic Index in its expanded form*

Clean

Organisms in order of tendency to be absent as the degree of pollution increases

		Total number of groups present[c]									
'Indicators'	Species richness				Biotic indices					
		0–1	2–5	6–10	11–15	16–20	21–25	26–30	31–35	36–40	41–45
Plecoptera nymphs present (Stoneflies)	More than one species	—	7	8	9	10	11	12	13	14	15
	One species only	—	6	7	8	9	10	11	12	13	14
Ephemeroptera nymphs	More than one species[a]	—	6	7	8	9	10	11	12	13	14
(Mayflies)	One species only[a]	—	5	6	7	8	9	10	11	12	13
Trichoptera larvae present	More than one species[b]	—	5	6	7	8	9	10	11	12	13
(Caddis flies)	Ome species only[b]	4	4	5	6	7	8	9	10	11	12
Gammarus present (Crustacean)	All above species absent	3	4	5	6	7	8	9	10	11	12
Asellus present (Crustacean)	All above species absent	2	3	4	5	6	7	8	9	10	11
Tubificid worms and/or Red Chironomid larvae present	All above species absent	1	2	3	4	5	6	7	8	9	10
All above types absent	Some organisms such as *Eristalis tenax* not requiring dissolved oxygen may be present	0	1	2	—	—	—	—	—	—	—

Polluted

[a] *Baetis rhodani* (Ephem.) not counted.
[b] *Baetis rhodani* is counted in this section for the purpose of classification.
[c] The term 'group' refers to the limit of identification without resorting to excessive keying-out.

Groups:
Each known species of Platyhelminthes (flatworms)
Annelids (worms excluding genus *Nais*)
Genus *Nais* (worms)
Each known species of Hirodinea (leeches)
Each known species of Mullusca (snails)
Each known species of Crustacea (Hog Louse, shrimps)
Each known species of Plecoptera (Stonefly)
Each known genus of Ephemeroptera
 (Mayfly, excluding *Baetis rhodani*)

Baetis rhodani (Mayfly)
Each family of Trichoptera (Caddis Fly)
Each species of Neuroptera larvae (Alder Fly)
Family Chironomidae (midge larvae except *Chironomus thunni*)
Chironomus thunni (Blood Worms)
Family Simulidae (Black Fly larvae)
Each known species of other fly larvae
Each known species of Coleoptera (beetles and beetle larvae)
Each known species of Hydracarina (water mites)

Modified from Hawkes 1979 after Woodiwiss.

which increases the 'sensitivity range' of the index towards the 'clean' end. There is a subjective choice of taxonomic groups and so therefore different sets of groups could be chosen for different river systems if it was considered that different types of rivers could not be compared using one form of the index. The main objective of this index is to monitor changes over time and not to compare index values between river systems but even so changes in the index value tell us little about the magnitude of the change in the biotic community. This index does not make use of changes in proportional species composition and such changes could occur as a result of changes in water quality.

More recently, the Trent Biotic Index was slightly modified by Chandler (1970) who added a semi-quantitative component (or weighting). This was done by Chandler in order to provide a scoring system (rather than an index) which would be dependent on relative abundance of groups present. Numbers of each macro-invertebrate group are recorded from five minute sampling periods. In Table 8.4 it can be seen that each abundant intolerant species would gain a high score but an abundant tolerant species would attract a very low score. In this system a score is allocated to each taxonomic group and the score for the sample is then derived by summing the individual group scores (Table 8.4). The Chandler Biotic Score can be modified to give the Average Chandler Biotic Score by dividing the Chandler Biotic Score by the number of species, species groups or genera used in the calculation (Murphy 1978). Similarly, the BMWP Score (see above) divided by the number of taxa gives an Average Score Per Taxon (ASPT).

Compared to the Trent Biotic Index, the Chandler Index quantifies sensitivity to organic pollution of macro-invertebrates by producing more detailed lists of species which have been ranked on their sensitivity to diminishing water quality. It is a subjective ranking and could be modified according to conditions and needs. In theory there is no upper limit for Chandler Biotic scores but values would rarely exceed 3000 and polluted streams have values up to about 300 (Balloch, Davies & Jones 1976).

The use of qualitative indices such as the Trent Biotic Index has been compared with quantitative indices by Murphy (1978) in an analysis of water quality from three rivers in Wales. Fig. 8.4 shows the temporal and spatial variation in the indices and it is clear that marked temporal variations in the Shannon–Wiener and Margalef indices mask any spatial patterns which may exist. Murphy concluded that for relatively un-polluted rivers, qualitative indices give a more consistent spatial dis-crimination between sites and that the average Chandler Biotic Score would seem to be at least affected by wide ranges in physical conditions.

Much could be said about the advantages and disadvantages of using

Table 8.4. *The Chandler Biotic Index*

		Increasing abundance				
		P	F	C	A	V
		Points scored				
Each species of	*Planaria alpina* Taenopterygidae Perlidae, Perlodidae Isoperlidae, Chloroperlidae	90	94	98	99	100
Each species of	Leuctridae, Capniidae Nemouridae (exd. *Amphinemura*)	84	89	94	97	98
Each species of	Ephemeroptera (exd. *Baetis*)	79	84	90	94	97
Each species of	Cased caddis, Megaloptera	75	80	86	91	94
Each species of	*Ancylus*	70	75	82	87	91
—	*Rhyacophila* (Trichloptera)	65	70	77	83	88
Genera of	*Dicranota, Limnophora*	60	65	72	78	84
Genera of	*Simulium*	56	61	67	73	75
Genera of	Coleoptera, Nematoda	51	55	61	66	72
—	*Amphinemura* (Plecoptera)	47	50	54	58	63
—	*Baetis* (Ephemeroptera)	44	46	48	50	52
—	*Gammarus*	40	40	40	40	40
Each species of	Uncased caddis (exd. *Rhyacophila*)	38	36	35	33	31
Each species of	Tricladida (exd. *P. alpina*)	35	33	31	29	25
Genera of	Hydracarina	32	30	28	25	21
Each species of	Mollusca (exd. *Ancylus*)	30	28	25	22	18
—	Chironomids (exd. *C. riparius*)	28	25	21	18	15
Each species of	*Glossiphonia*	26	23	20	16	13
Each species of	*Axellus*	25	22	18	14	10
Each species of	Leech (exd. *Glossiphonia*)	24	20	16	12	8
—	*Haemopsis*	23	19	15	10	7
—	*Tubifex* sp.	22	18	13	12	9
—	*Chironomus riparius*	21	17	12	7	4
—	*Nais*	20	16	10	6	2
Each species of	Air breathing species	19	15	9	5	1
	No animal life			0		

Levels of abundance in 'Score System'

Level	Nos. per 5 min sample
P = present	1–2
F = few	3–10
C = common	11–50
A = abundant	51–100
V = very abundant	100

Modified from Hawkes 1979.

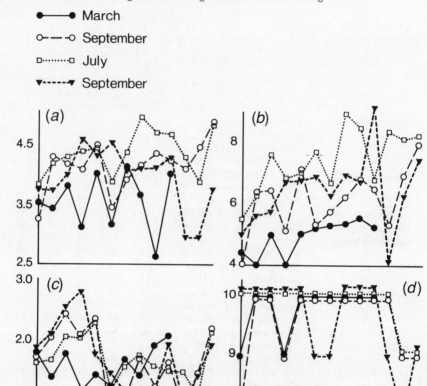

● ——— ● March
○ — — ○ September
□ ········ □ July
▼ ----- ▼ September

Fig. 8.4. Seasonal differences in various indices for the River Wye. Redrawn from Murphy (1978).

biotic indices as a basis for monitoring change in ecological systems. For example, the use of these biotic indices provides a quick, easy and cheap method of classifying certain characteristics of polluted water on a linear scale and also provides a basis for monitoring effects of pollution, but only organic pollution. That is, the biotic indices described above would not be suitable, without modification, for monitoring effects of certain kinds of industrial pollution where heavy metals were predominant.

Time consuming identification to species level can be avoided with the use of these indices. However, not all the taxonomic groups used for calculating the index may be suitable as detectors of change or be sensitive to all kinds of pollutants. Clearly there needs to be a judicious selection of the 'groups'. The species diversity or numbers of individuals is not recorded and in reality these biotic indices tell us little about community structure. One other potential disadvantage is that, depending on the sampling methods employed, natural changes in relative abundance of groups may be difficult to detect. For example, even in the case of a most basic sampling method, that of the kicking technique for sampling stream bottom fauna, duration and extent of kicking can result in a less than representative sample (Frost, Huni & Kershaw 1971). Some applications of biotic indices with reference to water pollution are described in Chapter 12.

Development of biotic indices

This brief and to some extent critical account of a few biotic indices may give the impression that biotic indices are simple and easy to develop. Nothing could be further from reality. In other words, this brief account does not take into consideration the many years of surveys and research which have led finally to the development and application of these biotic indices. Although the concept of an indicator species (Chapter 5) is quite simple, it is no simple task to transform that concept into practice. Once an idea for an index has been discussed, there then comes the need to consider which taxa to sample, which methods of sampling to use, the timing of sampling and the choice of sampling equipment. Plans on paper all too often turn out to be too ambitious or not cost-effective (often too much sampling is undertaken) when the pilot study has been completed. Results from the pilot study can provide a very useful basis on which to modify the definitive monitoring programme (see Fig. 10.1).

Temporal and spatial patterns of distribution of various taxa being sampled may demand that preliminary studies be undertaken for at least one year, leading to a further time period required for analysis and interpretation of data. Several years of modified sampling may then be required before it is possible to distinguish natural changes from those changes brought about by man's activities.

Development and use of biotic indices requires an interdisciplinary approach and therefore good communication between those disciplines is necessary. That need for good communication is important at both the research and development stage as well as at the interpretation stage, particularly when ecological concepts may have to be considered along with the planning and developments undertaken by local authorities and by industry. More comments about the importance and value of good communication are to be found in Chapter 10.

References

Abel, P. D. (1989). *Water Pollution Biology*, Ellis Horwood.

Balloch, D., Davies, C. E. & Jones, F. H. (1976). Biological assessment of water quality in three British Rivers: the North Esk (Scotland), The Ivel (England) and the Taf (Wales). *Water Pollution Control, London*, **75**, 92–110.

Cairns, J. (1979). Biological Monitoring – concept and scope. In *Environmental Biomonitoring, Assessment, Prediction, and Management – Certain Case Studies and Related Quantitative Issues*, ed. J. Cairns, G. P. Patil & W. E. Waters, pp. 3–20. Fairland, International Co-operative Publishing House.

Chandler, J. R. (1970). A biological approach to water quality management. *Journal of Water Pollution Control*, **69**, 415–22.

Dacre, J. C. & Scott, D. (1973). Effects of dieldrin on brown trout in field and laboratory studies. *N.Z. Journal of Marine and Freshwater Research*, **7**, 235–46.

Frost, S., Huni, A. & Kershaw, W. E. (1971). Evaluation of a kicking technique for sampling stream bottom fauna. *Canadian Journal of Zoology*, **49**, 167–73.

Hawkes, H. A. (1979). Invertebrates as indicators of river water quality. In *Biological Indicators of Water Quality*, eds. A. James & L. Evison, pp. 2.1–2.45, Chichester, New York, John Wiley.

Hellawell, J. M. (1978). *Biological Surveillance of Rivers*, Stevenage, NERC, WRC.

Hellawell, J. M. (1986). *Biological Indicators of Freshwater Pollution and Environmental Management*. London, New York, Elsevier Applied Science.

Hynes, H. B. N. (1960). *The Biology of Polluted Waters*. Liverpool University Press.

Inhaber, H. (1976). *Environmental Indices*, New York, London, Wiley.

James, A. & Evison, L. (1979). *Biological Indicators of Water Quality*. New York, London, Wiley.

Kreisel, W. E. (1984). Representation of the environmental quality profile of a metropolitan area. *Environmental Monitoring and Assessment*, **4**, 15–33.

Linstone, H. A. & Turoff, M. (1975). *The Delphi Method: Techniques and Applications*, Addison-Wesley.

Murphy, P. M. (1978). The Temporal Variability in Biotic Indices. *Environmental Pollution*, **17**, 227–36.

National Water Council (1981). *River Quality: the 1980 Survey and Future Outlook*, London, National Water Council.

Perring, F. H. & Farrell, L. (1983). *British Red Data Books, 1, Vascular Plants*, 2nd ed. Nettleham, RSNC.

Ribe, R. G. (1986). A Test of Uniqueness and Diversity Visual Assessment Factors using Judgement-Independent Measures. *Landscape Research*, 11, 13–18.

Richey, J. S., Mar, B. W. & Horner, R. R. (1985). The Delphi Technique in environmental assessment I. Implementation and effectiveness. *Journal of Environmental Management*, 21, 135–46.

Sackman, H. (1975). *Delphi Critique*, Lexington Books.

Scott, D. (1989). Biological Monitoring of Rivers. In *Proceedings of a Symposium on Environmental Monitoring in New Zealand with Emphasis on Protected Natural Areas*, ed. B. Craig, Wellington, Department of Conservation.

Sladecek, V. (1973). The reality of Three British Biotic Indices. *Water Research*, 7, 95–1002.

Sladecek, V. (1979). Continental systems for the assessment of river water quality. In *Biological Indicators of Water Quality*, eds. A. James & L. Evison, pp. 3.1–3.32. Chichester, New York, John Wiley.

Spellerberg, I. F. (1981). *Ecological Evaluation*, London, Edward Arnold.

Thomas, W. A. (1972). *Indicators of Environmental Quality*. New York, London, Plenum Press.

Tips, W. E. J. & Savasdisara, T. (1986). The Influence of the Environmental Background of Subjects on their Landscape Preference Evaluation. *Landscape and Urban Planning*, 13, 125–33.

Williams, N. V. & Dussart, G. B. J. (1976). A field course survey of three English river systems. *Journal of Biological Education*, 10, 4–14.

Woodiwiss, F. S. (1964). The Biological System of stream classification used by the Trent River Board. *Chemistry and Industry*, 443–7.

9

Biological variables, processes and ecosystems

Introduction

THERE ARE MANY DIFFICULTIES associated with monitoring ecosystems, not only because of the inherent complexities of ecosystems and
species interactions but because of the logistics and costs required to
support long-term programmes. These problems have been well described
in reports on long-term ecological measurements for the National Science
Foundation (NSF 1977, 1978). In 1977, a group of scientists met at Woods
Hole, Massachusetts, to discuss fundamental issues concerning long-term
ecological issues and not surprisingly they identified the need for special
funding, the need for specially adapted organizations to oversee the long-
term measurements, and that there is a definite need for such measurements. A pilot program for long-term observations and study of ecosystems was also suggested and this pilot programme was considered in
more detail at the second conference also held at Woods Hole.

These reports and other developments such as those initiated at the
Cary Conference (mentioned on p. 8) have focused attention on the fact
that there have been few long-term ecosystem monitoring programmes and
of the few that have been described in the literature, not all commenced as
monitoring programmes. If there are logistic and support costs, then why
bother to attempt ecosystem monitoring, particularly if it should be long-
term? Four reasons come immediately to mind.

1. The processes of many ecosystems have not been well researched and
   monitoring programmes could provide basic ecological knowledge.
2. Management of ecosystems, if it is to be effective, requires a baseline
   which can only come from ecosystem monitoring.
3. Man-made perturbations on the world's ecosystems have long-term
   effects, some synergistic and some cumulative and so therefore it
   follows that long-term studies are required.

4. The data from long-term studies can be a basis for early detection of potentially harmful effects on components of ecosystems.

Interestingly enough, the reports to the National Science Foundation for long-term observations of ecosystems made similar recommendations with regard to the data that could be provided and the nature of the measurements, viz. the data should provide: 1, cyclic changes; 2, detect time lags in ecosystem responses to outside influences; 3, test ecological theories concerning stability, community structure and system development; and 4, act as sensitive indicators of ecological change.

Monitoring at different levels of biological organization

It seems that much biological monitoring has previously been undertaken at the species level or at least at a simple level of organization. Monitoring at the species level is attractive because of lower costs yet much doubt has been expressed at the reliability of using single species to predict responses at higher levels of organization. In response to the often made claim that no significant ecological disasters have occurred on the basis of using single species to determine safe levels of impacts, Cairns (1984) has argued that it is not surprising that adverse effects have not been noted because the claim has not been extensively investigated.

The arguments for not undertaking biological monitoring at a high level of organization are the costs and possibly the limitations of practical facilities in the laboratory. It may, for example, be impractical to undertake laboratory-based investigations on the simultaneous and synergistic effects of many kinds of pollutants or impacts. Nevertheless, it can be argued that because single-species monitoring disregards interactions between species, the effects of environmental perturbations can only be realistically undertaken at an ecosystem level.

The immediate cost advantages of monitoring at the simple level of biological organization may well be obvious but such an approach demands much confidence in the ecological validity of monitoring at a simple level, particularly if short-term results are used to predict long-term consequences at higher levels of organization. Alleviating the unforeseen long-term effects of impacts on ecosystems could prove to be far more costly than the costs of carefully planned ecosystem monitoring.

Localities for ecosystem monitoring

The locality for long-term ecosystem monitoring needs to be secure and perturbations should be minimized. There are now possibly no localities in the world (apart from some at the bottom of the deepest oceans) which have not been affected by man but areas for long-term monitoring sites could be located where impacts are minimal òr if present may be incorporated in the monitoring programme. A secure location is equally as important as a location which has been the subject of previous recording and research. Some protected areas such as nature reserves or national parks fulfil these requirements and may therefore be suitable locations for long-term monitoring programmes (more comments on this on p. 250).

Perhaps because long-term field studies are uncommon there seems to have been very little research on effective, permanent marking of transects, quadrats and other data collection sites. The value of some long-term field studies based on permanent localities has diminished because of difficulties of precisely locating where data had previously been collected.

The recommendations in the National Science Foundation reports (NSF 1977, 1978), gave priority to sites with the following characteristics.

1. Sites where productive and useful short-term research has already been conducted.
2. Sites representative of major ecosystems and populations.
3. Sites considered important to major ecological issues.
4. Sites in which active scientists have shown interest.
5. Sites protected and accessible for research purposes.
6. Pristine ecosystems where crucial variables can be monitored.
7. Sites that are not pristine, but are unaffected by undesirable impacts and meet the other criteria.

Aquatic ecosystems

The acidification of inland waters throughout northern Europe and in Scandinavia has been given much publicity in recent years. Despite widespread public interest in 'acid rain' and its effects, we should perhaps remind ourselves that low-pH precipitation is not a new phenomenon. The Industrial Revolution (mid-18th century) was a major contributor to low-pH precipitation. Acidity of precipitation has increased about four-fold in northeastern USA since 1900. Records of 'acid rain' in Europe have been available now for more than 30 years (Fig. 9.1). The broad effects of diminished pH levels of inland waters is now well known, usually leading to diminished fish size and species richness. Despite some very interesting retrospective monitoring of acidification of inland water (example on p. 106), much more research is required and an understanding of the biological processes involved can be advanced by the establishment of monitoring programmes and by way of experiments.

For example, the effects of eight years' gradual, manipulated acidification on a small lake ecosystem in Ontario was described in 1985 by

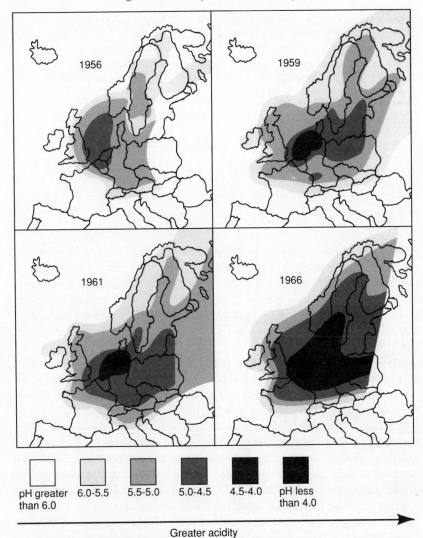

Fig. 9.1. The spread of acid rain in Europe from 1956–66. Redrawn from Ehrlich, Ehrlich & Holdren (1977) after Oden, S., Nederbordens forsurning-ett generellt hot mot ekosystemen. In Mysterud, I. (ed.) (1971) *Forurensning og biologisk mili-jovern*, pp. 63–98, Univeritetsforlaget, Oslo.

Table 9.1. *Biological parameters used to monitor a lake ecosystem during acidification over a period of seven years*

1. Species composition of epilimnion algae groups.
2. Phytoplankton productivity, biomass and diversity (Simpson's Index).
3. Epilimnion chlorophyll levels.
4. Density of dipterans emerging each year.
5. Percentage composition biomass of zooplankton groups.
6. 'Condition' of Trout.
7. Population levels of Crayfish (*Orconectes virilis*) and Sculpin (*Cottus cognatus*).
8. Size class structure of Minnows (*Pimephales promelas*) and Pearl Dace (*Semotilus margarita*).

From Schindler *et al.* 1985.

Schindler and his colleagues. In this experiment, which commenced with a two-year baseline survey, the pH was slowly decreased from a value of 6.8 to 5.0 and there were dramatic effects on the lake ecosystem which was studied over a ten year period. The selection of variables and processes chosen for that monitoring programme is of immediate interest (Table 9.1) because it is unusual to see such a wide range of variables and processes being employed.

The baseline survey was undertaken from 1974 to 1976. In 1976 there were few distinguishable chemical changes and little biological change. In 1977 the relative abundance of chrysophycean species (phytoplankton) declined slightly. The phytoplankton production, biomass and chlorophyll was within limits of natural variation for lakes in the area. In 1978 several 'key' organisms in the lake's food web were severely affected and primary production was slightly higher than in any previous year. In the next year, some algae species formed highly visible thick mats in littoral areas. In 1980, the condition of the lake Trout had declined, phytoplankton biomass had increased relative to the level in the previous year and an acidophilic diatom, previously rare in the lake, appeared in large numbers. The Trout continued to be affected; spawning behaviour changed in 1982 and in 1983 their condition was very poor and there was evidence of cannibalism amongst the Trout.

Many previous studies of acidification have relied largely on pH measurements and abundance of more common species of fish only. The report by Schindler *et al.* (1985) suggests that these are not sensitive, reliable indicators of early damage due to acidification. For example, twice-weekly pH measurements did not reveal the disappearance of 80 per cent of the alkalinity from the lake in the first year of acidification. They also suggested that most large fish are not sensitive indicators of early stages of acidification damage. The damage at lower trophic levels would, it was predicted, cause almost complete extinction of the Trout within a decade. If Trout are to be monitored then it needs to be more than just population size that is measured; other parameters include age-class structure and yield (yield is that part of production utilized by the consumer at a higher tropic level, or by man). Results from research in other Canadian lakes led Oglesby (1977) to conclude that fish yield has the advantage over traditional biotic and abiotic parameters partly because fish yield relates to socio-economic terms. This is perhaps a reminder that the objective of monitoring may not be solely for ecological reasons but there may be economic considerations as well.

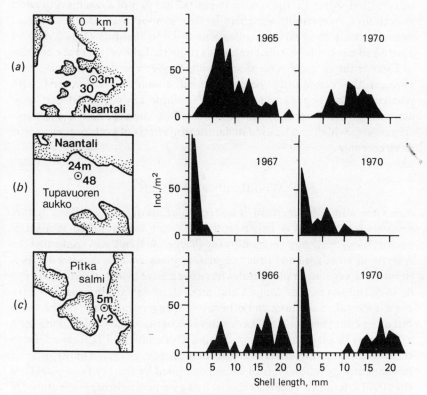

Fig. 9.2. Size-class distribution in populations of the mollusc *Macoma baltica* at three poluted sites off southwest Finland. (*a*) Decreasing population, no recruitment, (*b*) expanding population, intense recolonization prior to 1967, (*c*) aging population in 1966, newly started recruitment in 1969–70. Redrawn from Leppakoski 1979.

In Schindler's study, it was found that although the variables of phytoplankton production, changes in species diversity and richness were relatively insensitive to acidification, there were noticeable changes in species assemblages. For example, shifts from a large chrysophycean community to one where chlorophycean, cyanophycean, and peridinean species were often dominant in the phytoplankton. These shifts are characteristics of eutrophication.

The variables of species richness, species composition, species diversity and even biomass, when employed in isolation, would seem to have limited use in monitoring changes in ecosystems. Even age-class structure or size-class structure has it limitations despite the interesting observations that have been made (Fig. 9.2). The interpretation of these kind of data is difficult without information on abiotic variables and other biotic information. After all, the species in Fig. 9.2 are part of a community and a selection of community variables is best employed in monitoring eco-systems. But to be realistic, collections of large amounts of data may be costly and not easily justified unless, as suggested above, there are benefits to balance the costs. If ecosystem monitoring is to be done in detail, then perhaps the community being sampled should be considered with particular care. Leppakoski (1979) for example has suggested that large amounts of macrobenthos can be collected at low cost and is a suitable community which can be used in monitoring effects of pollution in aquatic environments.

Woodlands and forests

A review and assessment of long-term monitoring of woodland nature reserves by Peterken & Backmeroff (1988) not only resulted in a set of useful rules for ecosystem monitoring (Table 10.3) but also confirmed the scarcity of woodland and other vegetation plots used for monitoring. Few permanent vegetation plot studies have been established in central Europe or in Scandinavia. The longest and most thorough studies seem to have been undertaken on dune and other coastal vegetation in the Netherlands but other countries such as Sweden have recognized the value of long-term monitoring of plant community changes (Brakenhielm 1979) especially in relation to detecting effects of pollution on ecosystems. The Register of Permanent Vegetation Plots in Britain compiled by the ITE Ecological Data Unit (Hill & Radford 1986) has shown that a surprisingly diverse number of projects make use of an equally diverse range of vegetation (Table 9.2).

Peterken & Backmeroff (1988) described the history of five studies of changes in the structure and species composition of unmanaged woodland nature reserves in Britain, some based on transects and some on permanent

Table 9.2. *Types of vegetation monitored by permanent plots in Britain. These represent 63 projects being carried out by 27 people*

1. Lichens

2. Coastal sites
   Shingle heathland
   Shingle coast
   Dune grassland
   Saltmarsh
   Serpentine quarry

3. Calcareous grassland and scrub
   Railway and roadside verges
   Chalk grassland
   Juniper scrub (*Juniperus communis*)
   Scrub on limestone
   Calcareous flushes

4. Acid and neutral grassland and scrub
   Railway verges
   Acid grass
   Neutral grass
   Upland grass

5. Heathland, moorland and birch scrub
   Birch (*Betula*) on bog and moor
   Blanket bog
   Heathland and heather moor

6. Coniferous woodland and plantation forestry
   Sitka Spruce (*Picea abies*)
   Scots Pine (*Pinus sylvestris*)
   Pine on bog
   Amenity trees

7. Broadleaved woodland
   Acid oakwood
   Lowland coppice with standards
   Mixed woodland
   Calcareous ashwood
   Birch on bog and on moor

8. Aquatic vegetation
   Freshwater loch

Source: ITE Ecological Data Unit's Register of Permanent Vegetation Plots (Hill & Radford 1986).

quadrats. The two oldest transects were Lady Park Wood (Gwent and Gloucestershire) established in 1944 and Denny Wood (New Forest, Hampshire) established in 1954. The Denny Wood transect (Fig. 9.3), an acid oak–beech–holly wood, has been of particular interest in more recent years when there has been much concern expressed about the effects of acid precipitation. Occasional measurements of size-class distributions, growth rates and mortality data have shown that beech has regenerated much more than oak in recent times and many of the large beech but not oak have died. In an assessment of these trends, Manners & Edwards (1986) saw no reason to implicate atmospheric pollution and concluded that perodic drought (especially in 1976) aided by attacks of honey fungus has been largely responsible for a disintegrating wood. In the absence of more detailed data and synoptic recording of abiotic factors, it is reasonable to draw only general conclusions. A more detailed monitoring programme could include nutrient cycling measurements because the sensitivity of nutrient cycling to physical and chemical perturbations has been well demonstrated in other studies (see, for example, Ausmus 1984). With over three decades of records already available and at a time when effects of pollution on ecosystems need more careful monitoring, there is a convincing argument for the Denny transect and other woodland studies to be continued but always the costs of data recording, storage and analysis have to be considered.

Addressing the problems of costs associated with interpretation, replication and variability, Hinds (1984) has reasoned that non-destructive data on energy transfer between trophic levels in forest ecosystems (and other ecosystems) is a cost-effective basis for long-term monitoring. At the Battelle Pacific Northwest Laboratories, USA, Hinds and colleagues have focused on two contrasting ecosystems, closed canopy forests and toxic materials in food webs leading to avian predators. To detect slow and subtle changes in coniferous forest ecosystems they looked for ecologically valid data which could be obtained in a cost-effective manner. Conifers subjected to pollutant stress tend to lose their older needles sooner than do unstressed trees and for this reason Hinds has advocated the use of needlefall as an indicator of stress conditions. Coupled with basic climatic data, forest litter would seem to be one useful parameter to be used in long-term studies of forest ecosystems but with so few studies underway the alternatives have by no means been fully investigated.

Fig. 9.3. Denny Wood in the New Forest, southern England. This is an acid mixed woodland (oak, beech, holly) which was designated an ecological reserve by the Forestry Commission in 1952. The post in the foreground of the lower photograph marks the corner of a permanent quadrat. Photograph by B. Lockyer.

Estuarine ecosystems

Of all ecosystems, estuaries have probably been most severely damaged over the longest time-scale. This is because so many estuaries throughout the world have been the centre of habitation and consequently high population densities occur around them with the resulting impacts of pollution in its various forms. A major part of the estuarine ecosystem are the coastal wetlands which, because of their high productivity, are particularly rich in wildlife and this has meant that the anthropogenic impacts on wetlands have resulted in great losses of wildlife. In an effort to try and prevent further degradation of wetland sites, an international response in the form of the RAMSAR Convention (The Convention on Wetlands of International Importance Especially as Waterfowl Habitat) has been made.

The RAMSAR Convention was signed in 1971 in the Iranian town of Ramsar following a series of international conferences and technical meetings held in the 1960s, under the auspices of the International Waterfowl Bureau. By 1984, there were 39 parties to the Convention and 294 wetland sites covering about 20 million ha (Lyster 1985). Clearly, such conventions can only be successful if there are legally binding obligations but nevertheless this Convention has drawn attention to the need to closely monitor wetlands including estuaries.

The physical nature and geography of coastal ecosystems presents difficulties for the collection of data for long-term monitoring programmes. The tides, and in some locations the inhospitable terrain, make it very difficult to establish permanent sites and make it difficult to sample or record the fauna and flora. Methods for monitoring the long-term ecological effects of industrial impacts on estuaries has progressed in a very exciting way from simple ground surveys (sometimes in conjunction with aerial photography) to the very sophisticated digital remote sensing. These improvements and developments in monitoring methodology have been brought about in part by the need to be more cost-effective, in part by the need to be more precise and in part by the advances in remote sensing technology.

A classic example of these developments has emerged from long-term monitoring studies of the effects of a major refinery (one of the largest in Europe) and an oil-fired electricity generating station on salt marshes in Southampton Water on the south coast of England (Dicks 1976, Dicks & Iball, 1981, Dicks & Levell 1989, Shears 1989). The marshes (Fig. 9.4) have

Fig. 9.4 Fawley salt marsh on the edge of Southampton Water. One view looks towards a refinery and the other towards an oil-fired electricity generating station. Photographs by B. Lockyer.

been affected by industrial effluent since 1953 and damage occurred up until 1971. By then an area of marsh about 1000 m by 600 m close to the outfalls had been affected (Dicks & Hartley 1982). At that time, the refinery commenced a programme of enhanced effluent control and subsequently there was extensive recovery of the salt marshes. The salt marsh species *Salicornia* and *Sueda* were primary colonizers (Fig. 9.5) later to be followed by *Spartina* (which helps to stabilize the mud with its extensive roots systems). Spread of *Spartina* was also given some assistance by a refinery funded plant transplanting programme to areas where natural colonization was slow.

Monitoring commenced with a twice-yearly ground vegetation mapping technique which used a simple subjective scale of abundance of salt marsh plant density (Dicks 1976, 1989). At that time the objective of the monitoring was to describe the damage to the salt marsh community and to identify the cause of the damage. Following the improvements in effluent management from 1970 onwards, a broader monitoring programme was established to include littoral and benthic animals (Dicks & Iball 1981). The abundance categories for salt marsh plants were also revised to include the following for each species: abundant (majority of plants less than 50 cm apart and often close to each other), common (individual plants between 50 cm and one metre apart), rare (individual plants more than one metre apart).

In general terms, this monitoring, which largely involved twice-yearly ground vegetation mapping, was successful and produced very good correlations between the spatial distribution of the plants and proximity to effluent streams. Extending the surveys to animal groups was valuable particularly as some oligochaete worms were found to be particularly sensitive to effluent levels and thus good pollution indicators. The information has been used by the refinery in its management of effluents and led not only to a decrease in effluent levels but to a better understanding of ecological monitoring techniques necessary to assess the impacts of a large refinery. However, these ground vegetation surveys are time consuming and difficult to carry out. Shears (1988, 1989) therefore explored the use of remote sensing techniques for monitoring these salt marshes.

Shears used airborne thematic mapper (ATM) remote sensing data in conjunction with field spectroscopy and plant surveys to monitor the effects of the refinery effluent on the salt marshes. The objective of this work was to detect differences in the vegetation community and to map the distribution of the main salt marsh plant species. The ground surveys were based on ten parallel transects, seated at 100 m intervals, across the salt marsh. Two transects were in an area unaffected by effluents, two were in

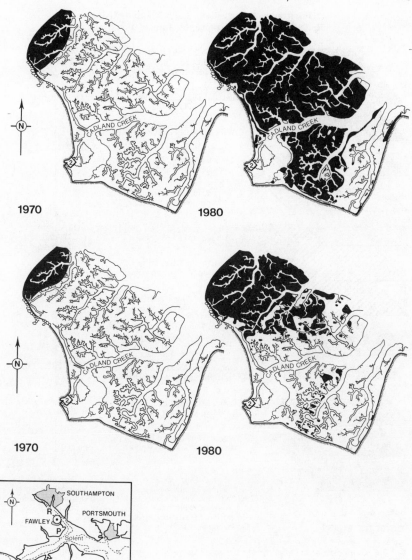

Fig. 9.5. Fawley salt marsh. Above, area of marsh covered by *Salicornia* in 1970 and 1980. Below, the same but for *Spartina*. R, refinery and P, electricity generating station. Redrawn from Dicks & Iball (1981).

Environmental Sensitivity Index Mapping Using
Remote Sensing and Geographic Information Systems

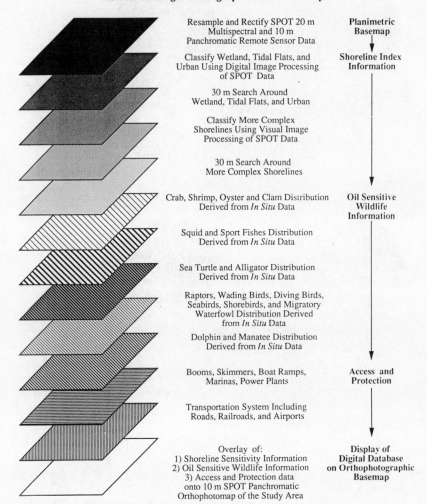

Resample and Rectify SPOT 20 m
Multispectral and 10 m
Panchromatic Remote Sensor Data

**Planimetric
Basemap**

Classify Wetland, Tidal Flats, and
Urban Using Digital Image Processing
of SPOT Data

**Shoreline Index
Information**

30 m Search Around
Wetland, Tidal Flats, and Urban

Classify More Complex
Shorelines Using Visual Image
Processing of SPOT Data

30 m Search Around
More Complex Shorelines

Crab, Shrimp, Oyster and Clam Distribution
Derived from *In Situ* Data

**Oil Sensitive
Wildlife
Information**

Squid and Sport Fishes Distribution
Derived from *In Situ* Data

Sea Turtle and Alligator Distribution
Derived from *In Situ* Data

Raptors, Wading Birds, Diving Birds,
Seabirds, Shorebirds, and Migratory
Waterfowl Distribution Derived
from *In Situ* Data

Dolphin and Manatee Distribution
Derived from *In Situ* Data

Booms, Skimmers, Boat Ramps,
Marinas, Power Plants

**Access and
Protection**

Transportation System Including
Roads, Railroads, and Airports

Overlay of:
1) Shoreline Sensitivity Information
2) Oil Sensitive Wildlife Information
3) Access and Protection data
onto 10 m SPOT Panchromatic
Orthophotomap of the Study Area

**Display of
Digital Database
on Orthophotographic
Basemap**

Fig. 9.6. Conceptual arrangement for environmental sensitivity mapping with remote sensing and GIS. This can be achieved if various kinds of information can be registered on a common planimetric basemap and then interrogated. From Jensen *et al.* (1990). Original figure kindly provided by John R. Jensen.

an area which had fully recovered and the remaining six were in areas still affected by the effluent. On each of the transects, ten sample points were chosen at random for collection of ground spectral data and plant biomass. Ground spectra were collected to help interpret the remote sensing imagery and were obtained using a NERC EPFS Spectron SE590 spectroradiometer. Plant biomass and species composition were also recorded from a 0.5 m² quadrat at each sampling point.

Computer image processing techniques were used to classify the marsh. An unsupervised classification, 'CLUSTER', where the computer automatically divides the data into spectrally distinct classes was found to give the most accurate results when compared with the ground survey; the resulting images could then be used to distinguish between several ecological zones on the salt marsh. Although the previous ground surveys by Dicks and his colleagues had identified different areas at different stages of recovery, it was only later that Shears was able to show possible uses of airborne remote sensing and ATM imagery to detect those parts of the salt marshes dominated by *Spartina* and *Salicornia* and thus provide an effective ecological monitoring technique.

Remote sensing using both aircraft and satellites has rapidly become a very sophisticated science with many applications. For example, in cases of large-scale applications of coastal monitoring, the satellite imagery from LANDSAT and SPOT (see Chapter 14) has been shown to offer advantages in terms of both costs and time (Hardisky, Gross & Klemas 1986). Remote sensing techniques and the use of Geographical Information Systems (GIS) which allow the merging of databases and mapping have also begun to provide an even more sophisticated basis for monitoring the effects of impacts on coastal ecosystems. For example Jensen *et al.* (1990), in their assessment of environmental sensitivity indices (ESIs), suggested that although information on oil-sensitive taxa on ESI maps can be useful, there is a limit to the amount of information that can be placed on a single map. Therefore, they suggest that information on the spatial distribution of oil-sensitive taxa is more effectively stored in a GIS database; individual files being created for each taxonomic group. A concept of creating databases using remote sensing and GIS technology is shown conceptually in Fig. 9.6.

References

Ausmus, B. (1984). An argument for ecosystem level monitoring. *Environmental Monitoring and Assessment*, **4**, 275–93.
Brakenhielm, S. (1979). Plant community changes as criteria of environmental changes. In *The Use of Ecological Variables in Environmental Monitoring*, ed. H. Hytteborn, pp. 73–80. The National Swedish Environment Protection Board, Report PM 1151.

Cairns, J. (1984). Are single species toxicity tests alone adequate for estimating environmental hazard? *Environmental Moniting and Assessment*, **4**, 259–73.

Dicks, B. (1976). The effects of refinery effluents: the case history of a saltmarsh. In *Marine Ecology and Oil Pollution*, ed. J. M. Baker, pp. 227–45, Barking, Applied Science Publishers.

Dicks, B.(1989). *Ecological Impacts of the Oil Industry, Proceedings of International Meeting Organized by the Institute of Petroleum and held in London in November 1987*. Chichester, New York, John Wiley, on behalf of The Institute of Petroleum.

Dicks, B. & Hartley, J. P. (1982). The effects of repeated small oil spillages and chronic discharges. *Philosophical Transactions of the Royal Society of London*, B**297**, 285–307.

Dicks, B. & Iball, K. (1981). Ten years of saltmarsh monitoring – the case history of a Southampton water saltmarsh and a changing refinery effluent discharge. *Proceedings of the 1981 Oil Spill Conference (Presentation, Behaviour, Control, Cleanup)*, pp. 361–74, Washington, D.C.

Dicks, B. & Levell, D. (1989). Refinery effluent discharges into Milford Haven and Southampton water. In *Ecological Impacts of the Oil Industry*, ed. B. Dicks, pp. 287–316. Institute of Petroleum, London, Wiley.

Ehrlich, P. R., Ehrlich, A. E. & Holdren, J. P. (1977). *Ecoscience, Populations, Resources, Environment*. San Francisco, W. H. Freeman.

Hardisky, M. A., Gross, M. F. & Klemas, V. (1986). Remote sensing of coastal wetlands, *BioScience*, **36**, 453–60.

Hill, M. O. & Radford, G. L. (1986). *Register of Permanent Vegetation Plots*. ITE, Abbots Ripton.

Hinds, W. T. (1984). Towards monitoring of long-term trends in terrestrial ecosystems. *Environmental Conservation*, **11**, 11–18.

Jensen, J. R., Ramsey, E. W., Holmes, J. M., Michael, J. E., Savitsky, B. & Davis, B. A. (1990). Environmental sensitivity index (ESI) mapping for oil spills using remote sensing and geographic information system technology. *International Journal of Geographical Information Systems*, A, 181–201.

Leppakoski, E. (1979). The use of zoobenthos in evaluating effects of pollution in brackish-water environments. In *The Use of Ecological Variables in Environmental Monitoring*, ed. H. Hytteborn, pp. 151–58. The National Swedish Environment Protection Board, report PM 1151.

Lyster, S. (1985). *International Wildlife Law*. Cambridge, Grotius Publications.

Manners, J. G. & Edwards, P. J. (1986). Death of old beech trees in the New Forest. *Hampshire Field Club and Archaeological Society, Proceedings*, **42**, 155–6.

NSF (1977). *Long-term Ecological Measurements. Report of a Conference, Woods Hole, Massachusetts, March 16–18, 1977*. National Science Foundation, Directorate for Biological, Behavioural and Social Sciences. Division of Environmental Biology.

NSF (1978). *A Pilot Program for Long-term Observation and Study of Ecosystems in the United States. A Report of a Second Conference on Long-term Ecological Measurements. Woods Hole, Massachusetts, February 6–10, 1978*. National Science Foundation, Directorate for Biological, Behavioural and Social Sciences. Division of Environmental Biology.

Oglesby, R. T. (1977). Fish yield as a monitoring parameter and its prediction for lakes. In *Biological Monitoring of Inland Fisheries*, ed. J. S. Alabaster, pp. 195–205. London, Applied Science Publishers.

Peterken, G. F. & Backmeroff, C. (1988). Long-term monitoring in unmanaged woodland nature reserves. *Research and Survey in Nature Conservation*, No. 9. Peterborough, Nature Conservancy Council.

Schindler, D. W., Mills, K. H., Malley, D. F., Findlay, D. L., Shearer, J. A., Davies, I. J., Turner, M. A., Linsey, G. A. & Cruikshank, D. R. (1985). Long-term ecosystem stress: the effects of years of experimental acidification on a small lake. *Science*, **228**, 1395–401.

Shears, J. R. (1988). The use of airborne thematic mapper imagery in monitoring saltmarsh vegetation and marsh recovery from oil refinery effluent – the case study of Fawley saltmarsh, Hampshire. *Proceedings of the NERC 1987 Airborne Campaign Workshop, 15th December 1988, Southampton*, pp. 99–119, NERC, Swindon.

Shears, J. R.(1989). The application of airborne remote sensing to the monitoring of coastal saltmarshes. In *Remote Sensing for Operational Applications*, ed. E. C. Barrett & K. A. Brown, pp. 371–9. Technical Contents of the 15th Annual Conference of the Remote Sensing Society, Remote Sensing Society, Nottingham.

Part C
MONITORING IN PRACTICE

10
Planning the monitoring

Introduction

THE DATA for monitoring and surveillance can be obtained either retrospectively or from a planned programme. In the case of some monitoring programmes the choice of variables is therefore limited because the data come from work which previously may not have been designed as a monitoring programme. Although a planned monitoring programme should be more sophisticated in terms of the choice of variables and in terms of the quality of the data, it does not necessarily follow that all retrospective monitoring is less successful than planned monitoring.

The example of the long-term monitoring of impacts of grazing Wildebeest in the Serengeti National Park (Belsky 1985), mentioned in Chapter 6, shows how good use can be made of information which was not necessarily collected for monitoring. Had the increases in population levels of the Wildebeest been predicted, then the monitoring would have been designed in a different manner. As it was, and in addition to field data collected by Belsky in 1982, the monitoring had to make use of reports of vegetation collected on previous occasions. Use was also made of aerial photographs of permanent plots established midway through the period of increased grazing as part of the Serengeti Ecological Monitoring Programme. In her research, however, Belsky (1985) was able to assess vegetation changes with accuracy and was also then able to suggest the best methods for a continuation of that monitoring programme.

There is no doubt at all that planning and design are an important part of any monitoring programme. That being the case, good use could be made of a conceptual plan or framework for the monitoring programme, whether it be monitoring environmental quality via organisms, or monitoring populations, habitats or ecosystems. Indeed a conceptual plan or framework is so important for both monitoring and environmental assessments (see Fig. 15.1) that there would appear to be a need for more research on conceptual plans. Although it would be difficult to establish a conceptual framework for all kinds of biological and ecological monitoring, the model suggested is of a very general nature and is meant to suggest good, basic practices (Fig. 10.1).

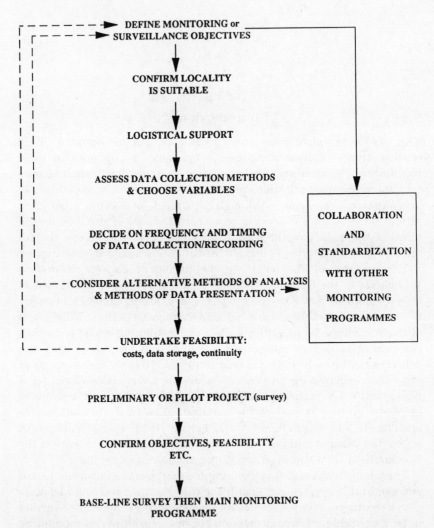

Fig. 10.1. Conceptual plan for a monitoring programme. Arrows indicate direction of steps that would be taken (solid lines) and revision of stages where necessary (broken lines).

The format and structure of a biological monitoring programme

The objectives

What seems to be an obvious first step, that of defining the objectives of a surveillance or monitoring programme, needs to be considered very carefully because all other components of the monitoring plan will be dependent on the objectives. After having written these comments, I was pleased to hear that a colleague in industry (Cowell 1978), when discussing monitoring as a management tool in industry, noted the following: 'It is the objectives ... that should discipline the sampling programmes that are undertaken, this seems to state the obvious, but I am alarmed at the number of monitoring schemes which I am asked to assess in which the objectives are either not stated at all or are so woolly that they are meaningless'.

A broad statement of objectives might initially be useful so long as there was subsequent further qualification of the objectives following a consideration of other aspects of the monitoring plan such as facilities for collection, storage and analysis of data. Indeed, realization of the full objectives may not be possible until baseline survey work has been undertaken. That is, without some biological and ecological information, defining the objectives becomes limited in its scope. However, a broad statement of the objectives might take the form of one of the following: to determine the abundance of bird populations in relation to agricultural practices; to assess effects of forest ride management on dispersion and dispersal of reptiles; to quantify characteristics of an aquatic ecosystem in relation to management of water quality; or to determine the effects of the development of an oil terminal on the physical and biological environment.

An example of the latter and a good example of planning objectives well in advance has been the monitoring undertaken by British Petroleum and the Field Studies Council at the Sullom Voe oil terminal, Shetland Islands (Westwood, Dunnet & Hiscock 1989). In 1974, oil exploration in the North Sea was well advanced and a decision was made to bring oil from the East Shetland Basin by pipeline to Sullom Voe. A Sullom Voe Advisory Group was established and planning for the monitoring of the effects of the development began before construction of the terminal commenced. A wide range of baseline studies was initiated and the monitoring programme has been audited by a specially established committee which in turn reports to the Shetland Oil Terminal Environmental Advisory Group. Monitoring has included work on the physical and biological components of the water, sediments and shores around Sullom Voe (further details in Chapter 15). The rocky-shore fieldwork has been undertaken by staff of the Field Studies Council (see mention of the Oil Pollution Research Unit on p. 81). The seabird populations were considered to be a special component of the environment and these have been monitored by a special working party. Such a large development inevitably brought the developers and the local community together and the resultant monitoring programme could not have been better planned. What is important is that planning and baseline studies took place, a lesson to be learnt for many other large and small developments undertaken by the oil industry.

Monitoring sites

Having defined the objectives, it is necessary to consider where the monitoring will take place and whether that locality would provide the most useful sources of the data (further comments on localities in Chapter 9). There is much to be said for using existing long-term monitoring sites, some of which have been described in Tables 3.1 and 9.2. The following are a few questions that may need to be considered. Are the localities suitable for monitoring, particularly in terms of the objectives? Will the locality be secure for the duration of the monitoring? Are the localities representative?

Some biological monitoring depends on data collected from precisely the same sites and permanent quadrats may have been established for this purpose. That being the case, not only should the locality be secure and representative but it is important that the sampling sites of the permanent quadrats can be located on each occasion the data is collected. Basic methods for marking permanent quadrats have been described on p. 88.

Arrangements for data collection and storage

Monitoring may be undertaken for long periods of time, during which the people initially involved in establishing the programme may change. Obviously there needs to be coordination of the programme so that records are not only deposited safely but are accessible and understandable to successive people involved in data collection. Standardization of methods as well as clear and precise descriptions of the methods used are of paramount importance for the continued success of any monitoring programme. Advances in information technology should be exploited.

While some biological monitoring programmes may require only one person or a few people to make regular data collections, other monitoring may require much larger data sets or data collected simultaneously from different regions of the country. It is for this reason that many monitoring programmes have drawn on the help of volunteers for the collection of data and the following are but a few examples. The Field Studies Council in Britain has drawn on the help of school children for data gathering (see p. 50). Some universities and colleges have made use of students in ensuring regular data collection; students at Southampton University have, for example, contributed to the monitoring at a woodland site (Denny Wood) in the New Forest (see p. 169). The Biological Records Centre, even from its earliest beginnings, relied very much on amateur help for collection of data. The British Society Lichen Mapping Scheme, the British Trust for Ornithology Common Birds Census and many other monitoring programmes rely on the work of amateur naturalists.

In recent years, two equally valuable programmes concerned with coasts have made use of support in the form of many, many volunteers. These two programmes contribute to a European-wide series of coastwatch projects in which ten countries play an active role. The N C C-based C O A S T W A T C H has been collecting basic information on the distribution and extent of Britain's coastal habitats and intensity of human activities. These basic data will provide a baseline for future monitoring of coastal habitats. The other coastal programme is the Norwich Union Coastwatch UK, a programme based at Farnborough College of Technology (Rees & Attrill 1989) and one which is helping to provide baseline environmental data in the fight to ensure pollution-free environments throughout Europe. Norwich Union Coastwatch UK is a very good example of a monitoring programme which not only relies on volunteer support but also makes very good use of this method for data collection (see address in Appendix II).

Collaboration and communication

The success of any monitoring programme depends firstly on good planning, secondly on the logistical support to continue the monitoring programme over the appropriate period of time and thirdly good coordination with other related programmes. It is interesting, with reference to the third point, that the US Federal Marine Pollution Plan (C O P R D M 1981) concluded that there was a need in marine monitoring for improved coordination among the existing monitoring programmes as well as more effective compatibility and communication. Steps necessary for good levels of coordination and management include the following (Wolfe & O'Connor 1986).

1. Develop an active inventory of regional and local monitoring activities to provide programme details for all users.

2. Establish uniform formats for a suite of selected monitoring parameters to facilitate accession and analysis of information.
3. Establish quality assurance systems to assure comparability of data programmes and regions.
4. Establish a national network of database management systems for storage, dissemination and analysis of data.

Selecting variables

There is a wide range of variables and processes, ranging from indicator species to productivity which can be used for biological monitoring and the use of many of these has been described in previous chapters. Ideally the choice of variables and processes should have a wholly ecological basis but logistic limitations (finance, time and effort) may override such considerations. Because data collection may be time consuming and expensive, methods for collection of data from the field or assemblage of data from other sources should therefore be considered along with the choice of parameters.

Simple variables may well be more cost-effective as well as meaningful in monitoring programmes and these aspects were included in recommendations for long-term ecological monitoring made to the National Science Foundation (NSF 1977, 1978), viz. measurement techniques should have the following characteristics.

1. Simplicity and reliability – so that studies made at different sites or times or by different investigators may be compared with confidence.
2. Stability – i.e. sites unlikely to change drastically over a period of decades or subject to rigorous intercomparison when techniques change.

To these we could add the need for standardized methods and a permanent record of the precise methodology which can be made available to any changes in staff involved. An example of the use of straightforward and simple variables is described in the Countryside Commission monitoring exercise on monitoring and management of wildlife habitats on farms (Matthews 1987). One of the many activities supported by the Countryside Commission as part of the Demonstrations Farm Project was the creation of a pond in a wet area of 'waste land'. The results of monitoring have been described by Usher (in Matthews 1987) and the monitoring was based on the use of plant species richness, abundance and composition (Table 10.1). On the basis of simple comparisons between 1979 and 1984 it was possible to give a fairly detailed but informative account of the changes which had taken place. Together with careful selection of variables and processes

Table 10.1. *A simple and effective basis for monitoring the plant species of a pond during establishment and management of wildlife habitats on demonstration farms. Here, results are taken from a study of a new pond established from wet waste land at Hopewell House, North Yorkshire, Britain. The plant species composition has changed and species richness has increased from 28 to 34*

Species	1979 abundance	1984 abundance
Agropyron repens (Common Couch)	•	f
Agrostis stolonifera (Creeping Bent)	•	a
Alopecurus geniculatus (Marsh Foxtail)	a	•
Arrhenatherum elatius (False Oat-grass)	•	a
Artemisia vulgaris (Mugwort)	•	r
Chamaenerion angustifolium (Rosebay Willowherb)	a	•
Cirsium arvense (Creeping Thistle)	f	a
Dactylis glomerata (Cock's-foot)	•	a
Deschampsia cespitosa (Tufted Hair-grass)	f	a
Epilobium hirsutum (Great Willowherb)	o	a
E. montanum (Broadleaved Willowherb)	•	o
E. palustra (Marsh Willowherb)	f	•
Equisetum arvense (Field Horsetail)	o	•
Galeopsis tetrahit (Common Hemp-nettle)	a	o
Gallium palustre (Common Marsh-bedstraw)	a	•
Glyceria declinata (Small Sweet-grass)	a	•
Heracleum sphondylium (Hogweed)	o	r
Holcus lanatus (Yorkshire fog)	•	o
Iris pseudacorus (Yellow Iris)	•	o
Juncus effusus (Soft Rush)	o	a

Table 10.1. (*cont.*)

Species	1979 abundance	1984 abundance
J. acutiflorus (Sharp-flowered Rush)	a	●
Lolium perenne (Perennial Rye-grass)	●	r
Lotus uliginosus (Great Bird's-foot Trefoil)	o	●
Odontites verna (Red Bartsia)	●	r
Phleum pratense (Timothy)	o	o
Phalaris arundinacea (Reed Canary-grass)	a	●
Plantago major (Greater Plantain)	r	●
Poa trivialis (Rough Meadow-grass)	r	●
Polygonum amphibium (Amphibious Bistort)	a	f
P. persicaria (Redshank)	●	o
Potentilla anserina (Silverweed)	a	r
Ranunculus repens (Creeping Buttercup)	f	a
R. ficaria (Lesser Celandine)	o	r
Rosa canina agg. (Wild Rose)	●	o
Rubus fruticosus agg. (Bramble)	●	o
Rumex crispus (Curled Dock)	o	●
R. obtusifolius (Broadleaved Dock)	r	r
Salix cinerea (Grey Willow)	●	o
S. fragilis (Crack Willow)	●	r
Scrophularia nodosa (Common Figwort)	●	o
Senecio jacobaea (Common Ragwort)	●	r
Solanum dulcamara (Bittersweet)	o	r

Table 10.1. *(cont.)*

Species	1979 abundance	1984 abundance
Sparganium sp. (Bur-reed)	●	?
Stachys sylvatica (Hedge Woundwort)	o	o
Taraxacum officinale agg. (Dandelion)	●	r
Trifolium pratense (Red Clover)	●	o
Urtica dioica (Common Nettle)	f	f
Veronica beccabunga (Brooklime)	f	●

Symbols: ● = absent, r = rare, o = occasional, f = frequent, a = abundant.

Taken from Matthews (1987) with kind permission of the Countryside Commission.

there may also be a need for judicious selection of timing and frequency of data recording. Where organisms are involved, phenological aspects must be taken into consideration as well as the spatial and temporal patterns of distribution (see p. 65).

Feasibility study

One of the greatest problems facing biological monitoring, especially long-term monitoring, has been the lack of financial and administrative support. Much ecological research has been based around short time periods of three or four years and opportunities to extend that research are rare. Although the advantages, benefits and need for more long-term monitoring projects have become obvious during the 1970s and 1980s, as a result of many pressing environmental issues, there is still a need to undertake structuring of the monitoring programme to make it cost-effective. Can structured monitoring be established? Wolfe *et al.* (1987) believe it can. They found general unanimity in the scientific literature on the importance of structuring the design of data collection but within a hypothesis-testing framework.

In our conceptual plan or framework (Fig. 10.1) we must therefore ensure that there will be logistical support for the monitoring programme. This means that we have to ensure that data will continue to be collected in a uniform manner and that methods of data storage and retrieval will be adequate. Continuity is of an essence and the longer the monitoring programme the greater the need to ensure that there is the support for that continuity.

A good strategy to adopt in any research is to think about the methods of analysis before the data are obtained because there is the possibility that data can be accumulated in a form which is not suitable for analysis. Such a strategy may therefore require a preliminary investigation.

Preliminary data gathering and baseline surveys

Planning biological monitoring can not take place without biological information and therefore data need to be assembled from published sources or from preliminary field surveys. Some of the basic ecological methods have already been outlined in Chapter 4 and attention was also drawn to a range of useful books on ecological methods such as Southwood (1966) and Gilbertson, Kent & Pyatt (1985).

The aim, at this stage, is to gather preliminary information about the biology of the area within the guidelines of the broad objectives. In its most simple form, this information could be in the form of species lists for selected taxa such as flowering plants, birds and spiders (see Appendix III) or it could be in the form of distribution of sensitive habitats and protected species (see p. 282 for comments about environmental sensitivity mapping). More detailed information might extend to the distribution and the population size and structure of the selected taxa.

To be able to quantify changes which may be the result of management, it is necessary to have a basis for comparison. As already emphasized, good baseline data, assembled as part of the monitoring plan, are an important prerequisite for successful biological monitoring. We can think in terms of baseline data as that information collected in the same place and on the same basis as subsequent data collection whereas reference data may have been collected previously. For example, data from monitoring of a population in relation to management could usefully be compared with populations trends observed elsewhere. This has been done with some data from the NCC Butterfly Monitoring Scheme (Pollard, Hall & Bibby 1986). In order to assess the results of butterfly habitat management, an index of abundance of each species of butterfly on a nature reserve can usefully be compared with regional trends (Fig. 10.2). In addition to these comparisons with regional trends, Pollard (1982) also made comparisons within a transect on the nature reserve to try and identify the cause of any departure from regional or national trends.

A baseline survey will also help in an assessment of which species will

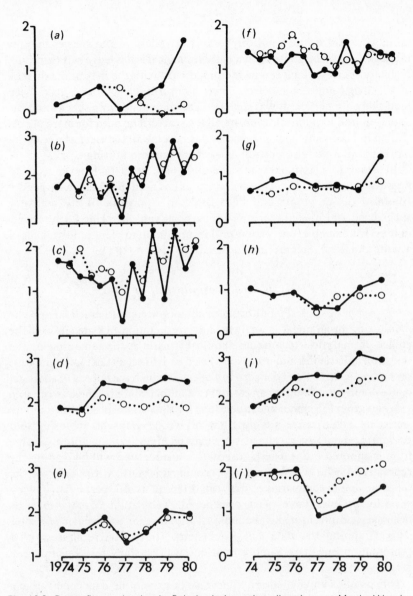

Fig. 10.2. Butterfly monitoring in Britain. Index values (logs.): — at Monks Wood compared with ... regional trends (R) or national trends (N). The starting point for regional trends is the Monks Wood Index value for 1974 and for national trends the Monks Wood index value for 1976. Monks Wood index values are excluded from regional data. (a) Grizzled Skipper (*Pyrgus malvae*); (b) Green-veined White (*Pieris napi*); (c) Peacock (*Inachis io*); (d) Gatekeeper or Hedge Brown (*Pyronia tithonus*); (e) Small Heath (*Coenonympha pamphilus*); (f) Brimstone (*Gonepteryx rhamni*); (g) Orange Tip (*Anthocharis cardamines*); (h) Speckled Wood (*Pararge aegeria*); (i) Meadow Brown (*Maniola jurtina*); (j) Ringlet (*Aphantopus hyperantus*). Redrawn from Pollard (1982).

be most useful in the monitoring exercise. Many of the species may be identified easily by those with a basic biological knowledge but there will be many species which can be identified only with the help and expertise of specialists and taxonomists. Some monitoring programmes have been modified only because it was soon realized that identifying species and the skill required for identification are all too scarce. The need for more people skilled in taxonomy and systematics has become all too clear as a result of more and more environmental assessment and monitoring work.

In order to emphasize the need for good skills in identification of organisms, a general list of guides and keys has been provided in Appendix III. Most of the guides and keys listed in Appendix III are not very specialized and indeed there is often a good case for using fairly simple guides. For example, one very useful introduction to plants and animals of a wide range of communities in Britain is Bishop (1973).

Analysis and presentation of data

An aspect of ecological monitoring easily forgotten until much later in the monitoring programme is analysis and presentation of data. Indeed the choice of analytical techniques should be made at the same time as the selection of variables and processes. The disadvantage of not following this procedure is that a lot of data can be collected which are not easily analysed or are not in a form which can easily be examined and presented in reports.

Apart from the common conventions for data analysis described in any statistical book, there are no right or wrong ways of analysing and presenting data. Several alternatives may be investigated before a suitable form is adopted but it may be useful to consider who will be reading the reports and who will be making recommendations on the basis of the reports. Simple summaries of data can often be useful, particularly when several variables have been considered. Tabulation of data may be necessary simply to make available all the data for assessment at a later date. Graphing the data can be a simple but effective first way of presentation but there is more than one way to show relationships in a graph.

One good example of alternative ways of presenting data comes from a study by Haefner (1970) in which mortality of Sand Shrimps was monitored in relation to the combined effects of temperature, salinity and oxygen. A summary of the data is shown in Table 10.2 and although not a large set of data, it is not immediately obvious which combination of variables causes least mortality. Haefner then puts the data in simple graph form, two examples of which are shown in Fig. 10.3, and the combined effects of temperature, salinity and oxygen become immediately obvious.

Table 10.2. *A summary of 4-day mortality data recorded for the Sand Shrimp*
Crangon septemspinosa *during exposure to low dissolved oxygen concentrations at*
12 different temperature–salinity combinations

Temperature (°C)	Salinity (%)	Percentage mortality		
			Female	
		Male	non-ovigerous	ovigerous
5.0	4.95	100	100	100
5.0	14.78	42.1	41.7	50.0
5.0	25.48	6.7	11.1	23.1
5.0	45.44	22.2	41.7	7.7
15.0	4.92	100	100	100
15.0	14.69	40.6	27.3	50
15.0	24.88	19.4	21.4	30.8
15.0	46.79	44.1	30.0	50.0
23.5	4.97	100	100	100
23.5	14.42	70.4	33.3	90.0
23.5	24.22	30.0	52.9	81.8
23.5	45.07	92.6	90.9	100

From Haefner (1970).

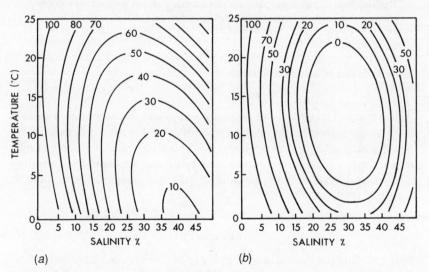

Fig. 10.3. Estimation of percentage mortality of ovigerous *Crangon septemspinosa*,
based on data from 12 combinations of salinity and temperature in (a) low concentrations of dissolved oxygen and (b) aerated water. Redrawn from Haefner
(1970).

Table 10.3. *A set of rules for ecosystem monitoring; with special reference to woodland ecosystems*

1. Any variable or process which can be readily measured and dated may be valuable in detecting changes in ecosystems.
2. Long-term monitoring must be supported by administrative continuity otherwise the programme may simply be overlooked or forgotten.
3. Facilities are required to ensure (i) survival of records and duplicate copies or records, (ii) markers locating the transect or quadrat and (iii) that the programme is known to exist.
4. Repetitive recording is obviously necessary and although it may not be necessary at regular intervals, further records should be taken after or prior to any formative events.
5. The monitoring locality should be inspected regularly (annually for a woodland) even if information is not collected.
6. Although objectives of the monitoring need to be defined, recording aims should be open-ended. The basic systematic record should be supplemented with casual adjuncts which have a habit of being valuable at a later date. This is because we don't know how the data could be applied in the future.
7. Simple variables and processes well recorded are more valuable than poorly recorded complex variables and processes. It's better to record something rather than nothing.
8. Representative records and replicates should be established if possible but even an unrepresentative sample may be valuable in the future analysis.
9. Regular analysis and preparation of reports, even at early stages in the monitoring help to improve the methods for data collection and help to refine the objective. These reports also serve as a reminder of the programme.

From Peterken & Backmeroff (1988).

The range of statistical and graphical computing packages now available on both main frames and mini computers provides an excellent basis for efficient data analysis and presentation. Those not familiar with the range of packages available are encouraged to attend appropriate computing courses and read around the subject (see, for example, Luff, Eyre & Rushton 1989).

Summary

Much of what has been said will appear to be obvious, but it is also obvious from the literature that many biological monitoring programmes have not been very successful because of a lack of planning, or lack of suitable and secure sites for collection of data or a poor selection of variables and processes used in the monitoring programme.

A few rules for the establishment and administering of biological monitoring programmes would seem to be a good idea and that is exactly what Peterken & Backmeroff (1988) have described in their assessment of attempts to monitor unmanaged woodland nature reserves (Table 10.3). These 'rules' which have been prompted by woodland ecosystem monitoring could equally apply to any other kind of biological monitoring.

Examples of a wide range of biological and ecological monitoring and surveillance programmes are described in the following six chapters. It is a personal selection and not all the examples are examples of good monitoring. The aim is to show the application (and sometimes misapplication) of the principles and concepts described in Part B.

References

Belsky, A. J. (1985). Long-term vegetation monitoring in the Serengeti National Park, Tanzania, *Journal of Applied Ecology*, 22, 449–60.

Bishop, O. N. (1973). *Natural Communities*, London, John Murray.

COPROM (1981). *National Marine Pollution Program Plan, Federal Plan for Ocean Pollution Research, Development and Monitoring, Fiscal Years 1981–1985*. September 1981. Interagency Committee on Ocean Pollution Research, Development and Monitoring, Washington, D.C.

Cowell, E. B. (1978). Ecological monitoring as a management tool in industry. *Ocean Management*, 4, 273–85.

Gilbertson, D. D., Kent, M. & Pyatt, F. B. (1985). *Practical Ecology for Geography and Biology Survey, Mapping and Data Analysis*. London, Hutchinson.

Haefner, P. A. (1970). The effect of low dissolved oxygen concentrations on the temperature–salinity tolerance of the sand shrimp, *Crangon septemspinosa*. *Physiological Zoology*, 43, 30–7.

Luff, M. L., Eyre, M. B. & Rushton, S. P. (1989). Classification and ordination of habitats of ground beetles (Coleoptera, Carabidae) in north-east England. *Journal of Biogeography*, 16, 121–30.

Matthews, R. (ed.) (1987). *Conservation Monitoring and Management. A Report on the Monitoring and Management of Wildlife Habitats on Demonstration Farms*. Cheltenham, Countryside Commission.

NSF (1977). *Long-term Ecological Measurements. Report of a Conference, Woods Hole, Massachusetts, March 16–18, 1977*. National Science Foundation Directorate for Biological, Behavioural and Social Sciences. Division of Environmental Biology.

NSF (1978). *A Pilot Program for Long-term Observation and Study of Ecosystems in the United States. Report of a Second Conference on Long-term Ecological Measurements. Woods Hole, Massachusetts, February 6–10, 1978*. National Science Foundation, Directorate for Biological, Behavioural and Social Sciences. Division of Environmental Biology.

Peterken, G. F. & Backmeroff, C. (1988). *Long-term Monitoring in Unmanaged Woodland Nature Reserves*. Research and Survey in Nature Conservation, 2, Nature Conservancy Council.

Pollard, E. (1982). Monitoring butterfly abundance in relation to the management of a nature reserve. *Biological Conservation*, 24, 317–28.

Pollard, E., Hall, M. L. & Bibby, T. J. (1986). *Monitoring the Abundance of Butterflies*. Research & Survey in Nature Conservation No. 2. Nature Conservancy Council, Peterborough.

Rees, G. & Attrill, C. (1989). *Norwich Union Coastwatch UK 1989 Report*. Farnborough, Farnborough College of Technology.

Southwood, T. R. E. (1966). *Ecological Methods, with Particular Reference to the Study of Insect Populations*. London, Methuen.

Westwood, S. S. C., Dunnet, G. M. & Hiscock, K. (1989). Monitoring the Sullom Voe Terminal. In *Ecological Impacts of the Oil Industry*, ed. B. Dicks, pp. 261–85. Chichester, New York, John Wiley.

Wolfe, D. A., Champ, M. A., Flemer, D. A. & Mearns, A. J. (1987). Long-term biological data sets: their role in research, monitoring, and management of estuarine and coastal marine systems. *Estuaries*, **10**, 181–93.

Wolfe, D. A. & O'Connor, J. S. (1986). Some limitations of indicators and their place in monitoring schemes. In *Oceans '86 Proceedings, Monitoring Strategies Symposium*, **3**, 878–84. Washington, Marine Technology Society.

11
Monitoring bird populations

Introduction

BIRDS are very often perceived as being more 'attractive' than other animal taxa and in general birds are more easily observed than some other animal groups. Possibly for these reasons, the distribution and abundance of birds has long been of interest to many people as well as organizations and societies throughout the world (see, for example, Durman 1976, Root 1988, Wild Bird Society of Japan 1982). More importantly, as far as monitoring is concerned, birds are useful indicators of land use change as is shown in responses of bird communities to agricultural landscapes and agricultural practices (O'Connor & Shrubb 1986). Agricultural chemicals are bound to be mentioned in any accounts about the biological effects of agriculture and in Chapter 1 we have already commented on Ratcliffe's classic study of the effects of DDT on egg shell thinning and on p. 11 there was mention of an avian pesticide monitoring based at Monks Wood. Here we are concerned more with changes in avian populations, some of which with careful analysis can be attributed to changes in the agricultural landscape and agricultural practices.

Sampling, recording and interpretation

Changes in bird population levels and distribution have been recorded for many years and we are fortunate in having long-term data on some species. A classic example is the population counts of Herons (*Ardea cinerea*) which is the longest census of any bird made in Britain and the data provide an excellent basis for analysis and interpretation (Fig. 11.1). There are of course regional variations in the data (Reynolds 1979) but on a national basis note that although the Heron population fluctuates, it does so within certain limits. Lack (1966) referred to this as 'restricted fluctuations' and also drew attention to the rapid recovery following severe winters. It is also interesting to note that rates of change in population levels do vary and tend to be greatest when levels of population are well below the mean. Could this mean that levels of reproduction in these Herons can be affected by population levels and/or population density?

The dramatic declines in Heron populations levels could, in general terms, be attributed to cold winters but such an explanation is an

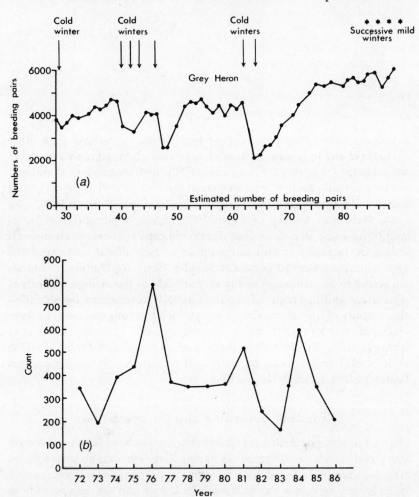

Fig. 11.1. Monitoring bird populations: (a) Grey Heron (*Ardea cinerea*) population levels in England and Wales (from Reynolds 1979 and kindly updated by the BTO); (b) Northern Harrier (*Circus cyaneus*) population levels (Titus *et al.* 1989).

oversimplification. In the absence of data from mortality studies we would need to be careful when we attempt to interpret these data because apparent changes in population levels could be attributed to a combination of ecological and observer factors. Causes of changes in population levels are as difficult to analyse as is the identification of patterns and trends. These Heron data and other examples of data from bird monitoring studies such as from monitoring studies on the Northern Harrier (*Circus cyaneus*) shown in Fig. 11.1 are so variable as to demonstrate that long-term studies are a prerequisite for the detection of trends. The Northern Harrier data and other aspects of monitoring raptors have been considered in some detail by Titus, Fuller & Ruos (1989) who conclude that observers must be alert to weather conditions which may affect counts.

As with counts of other animals, there will always be sampling errors and sources of inaccuracy which must be identified and taken into consideration in the final figures. Location of all birds on a count, and counting accuracy are two basic sources of error. In addition to these basic considerations (emphasized in Chapter 4), monitoring bird population levels requires decisions about sampling, time, duration and size of sampling.

Rothery, Wanless & Harris's (1988) analysis of counts of Guillemots is a good example of how to make these decisions when planning and analysing monitoring schemes. They counted samples of Guillemots in demarcated cliff plots in 27 colonies (ranging in size from 108 to 64 000) around Britain and Ireland and were especially interested in assessing validity of sample size or plot size as a reliable guide to total population size. Several underlying sources of variation in the counts, including simple day-to-day fluctuations and seasonal changes in populations, contributed to the coefficient of variation of about 10 per cent for plots of 200 birds or more. Analysis of detailed data from Isle of May suggested that the recommended model Guillemot monitoring scheme should consist of five plots of 200–300 birds counted on ten days. The timing of the counts should take into consideration the biology of the species at each colony. Since male birds depart at about the time when the young leave, Rothery *et al.* (1988) suggest therefore that counts are completed before the young leave, sometime in June. Clearly such monitoring programmes require a good knowledge of the ecology and biology of the species but Rothery *et al.* (1988) also noted that interpretation of changes observed in monitoring requires additional information from detailed demographic studies of the biology and ecology of the species.

Interpretation of a national change in avian species richness can be as equally interesting as interpreting data for a single species. For example, the data in Fig. 11.2 show changes in Britain's avian fauna from 1800 to

1972. Here we could usefully ask not only what are the reasons for this change but also how were the data collected and what are the limitations of these data? Does the apparent change in species richness reflect a change in the number of species and if so what could be the explanation? The apparent rapid increase (about five species per decade) could be the result of two activities, increased number of designated areas of protected habitats and increased interest in birds (that is increased observer effort), both activities that have long been promoted by The British Trust for Ornithology, the Royal Society for the Protection of Birds and other organizations.

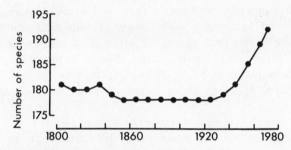

Fig. 11.2. Number of species (other than those introduced or reintroduced since 1800) breeding regularly in Britain and Ireland in each decade from 1800 to 1972. From Sharrock (1974).

Furthermore, general analysis of Britain's avian fauna can be undertaken by examination of changes in those bird species deemed to be 'failing' (species decreasing in numbers or whose ranges are contracting) or being 'successful' (increased numbers or expanded range). For this analysis, Sharrock (1974) looked at the status changes (increasing in numbers or expanding range, decreasing in numbers or diminished range, no change) for 129 species of birds in Britain over six unequal periods (Fig. 11.3). The persecuted species include Great Crested Grebes (feathers used on hats), Gannets and Gulls (collected for food), Woodpigeons (pests of crops), Goldfinches (trapped for cage birds) and Red-backed Shrikes (killed in the name of game preservation). The percentage of species 'failing' seems to decline up until 1939, after which it is followed by some increases, especially the wetland species. Reduction and damage to habitats has probably been the major contributing factor resulting in more wetland species 'failing'.

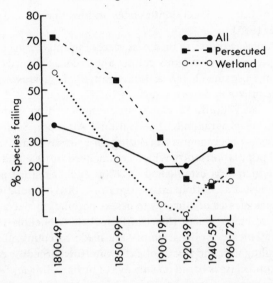

Fig. 11.3. Percentage of all bird species failing (see text) in each of six periods during 1800–1972. A = all species (129), P = persecuted species (935), W = wetland species (23). From Sharrock (1974).

Bird census and surveillance programmes

Bird census activities, both amateur and professional, range from local and occasional recording to the large and well-organized national programmes. Not surprisingly the different census programmes have different objectives and different methods, a subject which has been well debated in periodic meetings on bird census and studies (see, for example, Taylor, Fuller & Lack 1985). There is an intrinsic interest in bird census programmes, but in addition some bird species may usefully be used as indicators of the state of the environment, for example in monitoring marine ecosystems.

Well-organized long-term census programmes are a prerequisite for monitoring and the following are examples of census programmes which have provided valuable information for monitoring the distribution and abundance of birds. In assessing these examples, I find it useful to ask: what is the aim of the programme and how is the information collected and analysed? These questions provide a necessary basis for the subsequent interpretation of the data. Information on other census programmes can be found in some of the British Trust for Ornithology publications such as the

Proceedings of International Conferences on Bird Census and Atlas Work
(Taylor *et al.* 1985).

In North America there are bird nest recording and census programmes,
ringing programmes and various counts and records of birds throughout
the continent, organized by a wide range of groups. Government
organizations in North America such as the US Fish and Wildlife Service
and the Canadian Wildlife Service sponsor the Breeding Bird Survey and
also bird ringing programmes. The National Audubon Society sponsors
three bird census programmes: a breeding bird census established in 1937,
a winter bird population study which commenced in 1948 and a Christmas
Bird Count which was established in 1900 and has since proved very
popular amongst its participants. Migratory bird populations are the
subject of three classes of surveys to assess population size as a basis for
monitoring and management (Martin, Pospahala & Nichols 1979). In the
first class, statistically valid attempts are made to count all individuals
within a sampling unit and this applies to waterfowl breeding counts in the
USA. The second class is based on surveys of birds calling and/or observed
within a sampling unit; one species which is the subject of this class of
survey is the Mourning Dove (*Zenaida macroura*). The third class described
by Martin *et al.* (1979) is represented by total counts, within designated
sampling areas, of species such as wintering waterfowl.

There is a wide range of bird distribution and population recording,
census and surveillance activities in Britain (Table 11.1). For example the
British Trust for Ornithology (BTO) is the sole organization responsible for
bird ringing and is the major contributor to bird census and surveillance
work in Britain via the Common Birds Census, the Waterways Bird Survey,
the Nest Record Scheme and more recently by an Integrated Population
Monitoring programme which now embraces several programmes (Baillie
1990, Marchant *et al.* 1990). The BTO Atlas Project has recorded the
distribution of both breeding and wintering birds for many years
(Sharrock 1976) and the distribution maps of each bird species (based on
presence or absence in each 10 km sq of the National Ordnance Survey
Grid) can be used for baseline information in monitoring programmes. The
Game Conservancy administers the National Game Census, the aim of
which is to monitor the number of game shot and thus derive population
trends for species such as Pheasant, Red Grouse, Grey Partridge, Brown
Hare and Wood Pigeon. The National Game Census has continuous
records which were established in 1961 but also has records going back to
1880. The Wildfowl and Wetlands Trust participates in census of
wildfowl, both at an international and national level and the Royal Society
for the Protection of Birds has monitoring programmes including those
directed at threatened species.

Table 11.1. *Examples of some bird recording, census and surveillance programmes and activities in North America and Europe*

North America
Hawk Migration Association of North America

National Audubon Society
Breeding Bird Census (est. 1937)
Winter Bird Population Study (est. 1948)
Christmas Bird Count (est. 1900)

US Fish and Wildlife Service & Canadian Wildlife Service
Breeding Bird Survey (est. 1966)

Europe
British Trust for Ornithology

The BTO Integrated Population Monitoring Programme, incorporating:
(i) The Common Birds Census, CBC (est. 1961, farmland and woodland species)
(ii) Waterways Bird Survey, WBS (data since 1974 for riparian species)
(iii) Nest Record Scheme (est. 1939)
(iv) Constant Effort (mist-netting) Sites scheme
(v) Ringing Scheme

Heronies Census (data since 1928 with some counts commencing in 1909)
Garden Bird Feeding Survey
Birds of Estuaries Enquiry (est. 1969 with 'constant effort' sites established in 1985, shorebirds and wildfowl)

Centre de Recherches sur la Biologie des Populations D'Oiseaux (CRBPO), France
STOC (commenced in 1989 with ringing sites and point counts)
GISØ Oiseaux Marins, France (seabird survey)

Game Conservancy (Britain)
National Game Bird Census (est. 1961)

Lund University (Department of Ecology), Sweden
National Wildfowl Census (commenced 1959)

Nature Conservation Board (Sweden)
The Breeding Bird Census (part of the Programme of Environmental Quality Monitoring). Commenced 1970

RIN (Research Institute for Nature Management), The Netherlands
International wader counts, National goose counts, Mid-winter wildfowl counts in association with IWRB)

Royal Society for the Protection of Birds (Britain)
Surveillance of threatened species (e.g. Osprey *Pandion haliaetus* since 1954)
Beached Seabird Survey (est. 1971, but now discontinued)
Monitoring of some seabirds (since 1971)

Table 11.1. (*cont.*)

SOVON (Cooperating Organizations on Bird Census work in the Netherlands). Counting of landbird migrations. Common breeding bird census, Rare breeding bird project, Point transect counts of wintering birds

Swedish Ornithological Society (Sveriges Ornitologiska Forening)
The Winter Bird Census (commenced 1975/76)
Nest Box Monitoring (commenced 1980)

Wildfowl trust (Britain)
National Wildfowl Counts; Mid-winter Wildfowl Counts since 1946 (now in association with the International Waterfowl Research Bureau)

WWF – Nordrheinwestfalen (Germany)
Goose counts in FRG in association with IWRB

See Durman (1976) for a useful list of bird observatories in Britain and Ireland.

Data collection, analysis and interpretation

There are many methods of data collection for bird census studies. Recording (by sight and sound) species along a transect or within a defined habitat are commonly used techniques. There are many variables which may affect the results such as amount of time spent recording, rate at which the transect or habitat is walked, time of day or year, weather conditions and the experience of the recorder. Point Count techniques such as the Variable Circular Plot Technique (Reynolds, Scott & Nussbaum 1980) are popular with North American observers and provide a standardized technique for estimating bird numbers. Basically, the Variable Circular Plot Technique consists of making series of timed stops during which the distance at which each bird is first detected is estimated. For each species, an Effective Detection Distance is determined, beyond which detections drop off sharply. Numbers and densities are calculated from records within the Effective Detection Distance.

If one of the aims of the census is to prepare distributions maps, which may later then be used for monitoring changes in distribution and relative abundance, then many participants in many locations is a necessary prerequisite. Bird counts such as the Christmas Bird Count in the USA are so popular that there were 38 346 participants in a recent (1985–86) count. With the use of specific locations, each of which covers a 24 km radius, and a standard 8 hour recording time at each site, the Christmas Bird Count

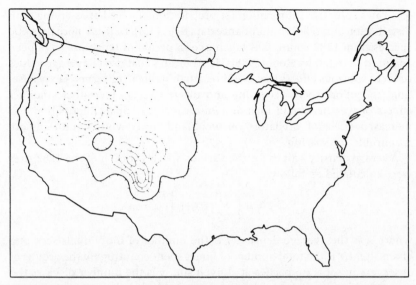

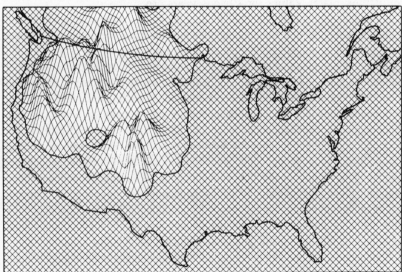

Fig. 11.4. Contour and three-dimensional maps of the winter distribution and abundance patterns of the Black-billed Magpie (*Pica pica*). The four contour intervals are 20%, 40%, 60% and 80% of the maximum abundance value used in the normalization process. For the Black-billed Magpie, this value is 18.30 individuals seen per party-hour. Reproduced from Root (1988) with kind permission of the author and Chicago University Press.

generates sufficient information for preparation of a bird atlas giving both distribution and relative abundance data (Fig. 11.4). Data on more than 600 species from 1282 count sites taken over a period of ten years have been analysed in detail by Root (1988). He advocates that use of average values reduces spurious effects that could be attributable to observer's expertise, time and effort spent recording and direct effects of weather. Indirect effects of weather could include diminished participant motivation and increased effect of predators on woodland birds when conditions are favourable for soaring.

Average density values for the various species at each of the 1282 sites were calculated as follows:

$$\bar{X} = \left[\sum_{i=1}^{y} (I/Hr)_i \right] \bigg/ y$$

where X is the average density, I is the number of individuals seen at a given site, Hr is the total number of hours spent counting by the groups of observers in separate parties at a given site, y is the number of years the count took place. Values for various years are summed.

For ease of computer plotting, the density values were averaged to range between zero and one for each species by dividing the average values at each site for a given species by the average value at the site with the maximum abundance or the value which was greater than 99 per cent of all the values for a given species (maximum abundance value). An example from Root (1988) clarifies this calculation. The highest value for the Common Raven was 6.67 individuals seen per party-hour while the next three highest abundances were 18.24, 15.76 and 13.02 individuals per party-hour. Instead of using the absolute maximum, Root uses the average value which was 99 per cent greater than all other values for a given species. This value for the Common Raven is 13.02 and the values higher than this were given a value of 1.0 for the computer mapping. An example of computer mapping for the Black-billed Magpie is shown in Fig. 11.4. This kind of mapping will no doubt become more sophisticated with advances in GIS and allow for more detailed analysis of data.

Bird census studies can provide information on patterns of distribution and relative abundance but usually not population size because, with the exception of some species with small populations, it is impossible to count all individuals. The British Trust for Ornithology's Common Birds Census (Marchant 1983, Marchant et al. 1990), commissioned by the Nature Conservancy Council, has tackled this problem in an interesting manner. The original objective of the Common Birds Census (CBC), which commenced in 1961 (the year before Silent Spring was published), was to monitor any adverse effects on bird populations that may have been caused by agricultural chemicals, especially pesticides. Today the CBC is designed to detect and measure population changes of various bird species (not only common species) in woodland and on farmland. The census is based on data collected mainly from 350 woodland and farmland plots which are visited on several occasions during March to July. The farmland plots can be of any type but must be at least 40 ha, and preferably 60 ha, in area. The woodland plots include all kinds of semi-natural vegetation, each at least 10 ha in area. Records of all birds heard or seen are plotted onto large-scale maps provided by the BTO.

The field data from the CBC are collated, recorded and analysed in a number of different standardized ways and the index of population change for a period of up to 30 years has proved especially popular when analysing trends in populations. The index of population change is relative to an arbitrarily chosen datum year where the index is set at 100. The datum year was previously 1966 but is now set at 1980 for most species (Fig. 11.5). It is important to note that data collection and analysis is highly

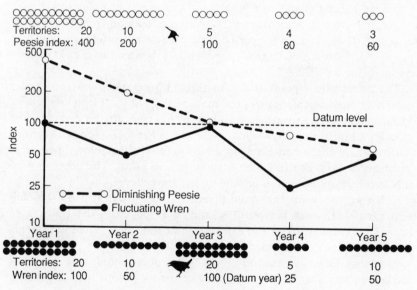

Fig. 11.5. The basis of the Common Birds Census Index is illustrated by population indices for two hypothetical species: the Diminishing Peesie *Vanellus disparator* (broken line) and the Fluctuating Wren *Troglodytes updownii* (continuous line) over a period of five years. The index is set to an arbitrary value of 100 (in this case year 3) and is a measure of a species change in abundance relative to the datum year. In some years these two hypothetical species have the same number of territories but the index value differs for each species, the Peesie having a higher value throughout despite the lower number of territories in years 3 to 5. From Marchant (1983).

standardized, so that any temporal trends observed are not the result of subtle changes arising out of differences in data collection or analytical methods that might occur over time.

The Common Birds Census Index and the Waterways Bird Survey Index are calculated by way of the chain method, that is data on the number of territories from each plot are paired with those from the same plots in the previous season. Such a method helps to eliminate variation which may be caused by census accuracy and observer turnover. The counts are then summed across all pairs to produce an overall estimate of percentage change. This estimate is then applied to the previous year's index value.

This method has provided a very useful basis for assessing and monitoring population change at a national level. With regard to interpretation of the index, it is interesting to consider the overall changes (which exclude regional differences) with respect to various species (Fig. 11.6). The explanations for changes in bird populations levels are many and include the effects of weather especially severe winter weather, changes in rainfall patterns and habitats at the overwintering locations of migrants, habitat changes in Britain and the effects of pollutants.

The following (Fig. 11.6) are a few examples selected from Marchant *et al.* (1990). These data cover a period of time when there was a severe winter in 1962–63 followed by no really hard winters then a series of mild winters commencing in 1972–73.

The palaearctic migrant, the Whitethroat (*Sylvia communis*) is a species associated with heathlands and commons with good scrubland vegetation. The dramatic decline during 1968 and 1969 has been attributed not to loss of habitat but to mortality when wintering in the Sahel Zone of Western Africa where there had been severe droughts (Winstanley, Spencer & Williamson 1974, Peach, Baillie & Underhill, in press). The population of this species now fluctuates around a new lower level.

The CBC Index for the Wren (*Troglodytes troglodytes*) is outstanding with a tenfold increase between 1963 and 1974, suggesting that this species has a unique ability to respond to favourable conditions. Winter weather is a major factor in this species' mortality.

Lapwings (*Vanellus vanellus*) were commonly found over much of lowland Britain but suffered a decline after the 1962–63 winter. In general, there seems to have been a decline in southern areas (due to changes in farming practices and drainage of damp meadows) but an increase in the north, possibly due to an ameliorating climate. Conservation of this species could usefully be aimed at establishing traditional wintering sites such as large, old, fertile pastures where organic farming is practised.

The Chiffchaff (*Phylloscopus collybita*) CBC Index shows a dramatic increase, an equally dramatic decline, then a fluctuation leading to an

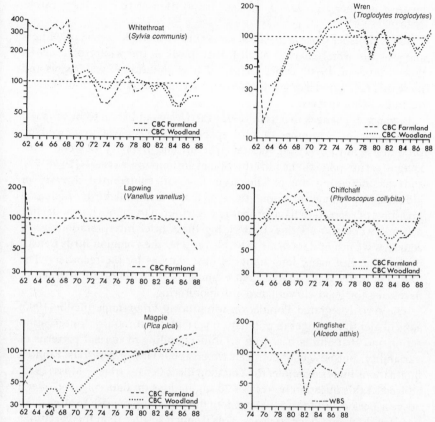

Fig. 11.6. Examples of BTO Common Birds Census and Waterways Bird Survey index graphs. Note that although these can be used for surveillance of species, there can not be numerical comparisons between species. Each index relates to the relationship of that year's population level to the datum year of 1980. From Marchant et al. (1990).

increase. The decline in the early 1970s is possibly a reflection of loss of habitat which is typically either old, mixed deciduous woodland with some tall trees but with a good undergrowth or shrub and scrubby woodland edges. Winter conditions may also be an important factor for this species.

Following persecution for protection of game earlier this century, the Magpie (*Pica pica*) has an increasing population in both farmland and woodland habitats. It appears that this species has been able to adapt to wooded landscapes in urban areas. There is some evidence to show that higher densities of Magpies in urban areas do occur where there is a greatest number and varied species composition of tree species (Tatner 1982).

The Kingfisher (*Alcedo atthis*) is especially vulnerable to severe winters and was the worst affected of all British birds in the winter of 1962–63 (Marchant *et al.* 1990). Other factors such as pollution of waterways and increased recreation along waterways have contributed to the longer-term decline of this species.

In general, changes in agricultural practices would seem to be of major importance for a number of bird species, a subject well reviewed by O'Connor & Shrubb (1986). Habitat structure and diversity (of woodlands, hedges, scrub, ponds), the introduction of autumn sown cereals (1974–77), and the increasing use of chemicals have all contributed directly or indirectly to change in populations of many farmland birds. Increased pressures of recreation have probably contributed to a more recent decline of some woodland species. Underlying these brief interpretations is, of course, a wealth of data made possible only by the Common Birds Census programme and many long hours of observations by the recorders. The CBC is in many ways an excellent example of the kind of infra-structure necessary for good surveillance and monitoring.

The BTO Integrated Population Monitoring Programme (Baillie 1990) aims to identify changes in population variables that require conservation action and that aim is based on an understanding of normal patterns of variability. But how confident can we be about lack of spurious, random fluctuations in, for example, the Common Birds Census Index? Moss (1985) addressed this question by way of 20 simulations for data collected over a 20-year period and found that random fluctuations gave rise to at most 25 per cent deviation over 20 years. This is much smaller than the variations in population levels that have been estimated using the Common Birds Census data. Moss also looked at the relationship between the index values (published in the journal *Bird Study*) and population density values for plots where the area of the plot is known. In some cases there was a high correlation between density of breeding pairs and the Common Birds Census Index, suggesting that the index does reflect actual density in an accurate manner. However, in one of the examples chosen, the Spotted Flycatcher (*Muscicapa striata*), there was no relationship between the index and the population density over a 20-year period as a whole. Moss attributed this to the fact that new plots coming into the Census on average held more Spotted Flycatchers than those plots dropping out and so the Index drifted downwards.

Other variables, such as weather conditions during population counts and the level of observer expertise, need to be considered when assessing data from census programmes. The Common Birds Census provides a rich

GREY PARTRIDGE

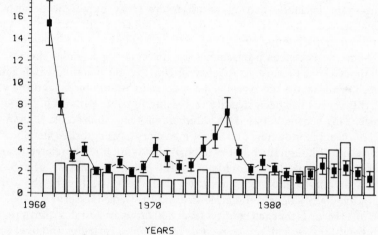

WOODCOCK

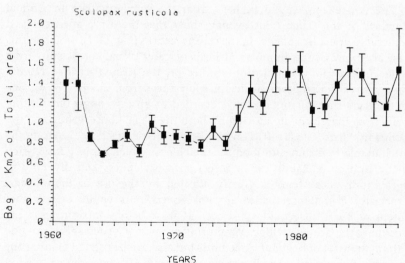

Fig. 11.7. Game Conservancy data for the Grey Partridge *Perdix perdix* and the Woodcock *Scolapax rusticola*. From Tapper & Cook (1988). Graphs kindly provided by S. Tapper.

source of data for analysis of these variables. Variations in weather conditions have been found to be too small to bias Common Birds Census results (O'Connor & Hicks, 1980) and whereas expertise was found to be an important variable, pairing of observers (very experienced with less experienced observer) was found to eliminate influence of observer differences (O'Connor 1981).

Some birds census programmes are directed at particular species. The National Game Census, established in 1961, is one example which aims to monitor the status of several game species, including birds, and the value of this census has been steadily increasing in value with each succeeding year. This census is based on numbers of game shot across Britain. For many years landowners and shoot managers have recorded details of the bags, and although the number of birds shot is not linearly related to actual bird density, it does provide a useful index to monitor year-to-year changes and trends. It is assumed that the reliability of the data is improved with increased sample size and the Game Conservancy looks for 10 to 20 records for each county. Over 500 farms and estates return records and of those over half return records for the Grey Partridge and over 80 per cent return records for the Woodcock. Census forms ask for exact totals for all game taken on the shoot in the preceeding 12 months as well as number of any game released.

The two examples shown here from the summary data in Annual Reviews of the Game Conservancy show that there are apparent and sometimes dramatic changes in the census returns (Fig. 11.7), which in this case are expressed as numbers of birds per area of habitat. Results from Game Conservancy monitoring studies on the Grey Partridge (*Perdix perdix*) have shown dramatic declines in returns from shoots. The decline in numbers shot mirrors declines recorded by spring counts of breeding pairs from long-term monitoring by both the Game Conservancy and the Common Birds Census (O'Connor 1985). These declines seem to be attributable to declines in food for the chicks, which require insect-rich diets (Hudson & Rands 1988). Insect species richness and diversity in agricultural ecosystems is greatly affected by the use of insecticides, especially if the insecticides are not selective. It would appear that intensive use of insecticides has been a major factor contributing to the decline in the Grey Partridge. Monitoring programmes such as this provide information not only about game birds but also can be used to monitor the impacts of chemicals on agricultural ecosystems. However, interpretation of data needs to be undertaken with care. For example, the apparent increase in Woodcock (*Scolopax rusticola*) may not reflect a true increase in numbers. The increase in the returns of birds shot is partly a reflection of the increase in Pheasant shooting. Most Woodcock are shot during Pheasant drives and the larger numbers of Woodcock shot may be biassed by this increase in hunting effort.

References

Baillie, S. R. (1990). Integrated population monitoring of breeding birds in Britain and Ireland. *Ibis*, **132**, 151–66.

Durman, R. (1976). *Bird Observatories in Britain and Ireland*. Berkhamsted, T. & A. D. Poyser.

Hudson, P. J. & Rands, M. R. (1988). *Ecology and Management of Gamebirds*, Oxford, Blackwell Scientific Publications.

Lack, D. (1966). *Population Studies of Birds*. Oxford, Clarendon Press.

Marchant, J. (1983). Common Bird Census Briefing, BTO *News*, **128**, 9.

Marchant, J. H., Hudson, R., Carter, S. P. & Whittington, P. A. (1990). *Population Trends in British Breeding Birds*, Tring, British Trust for Ornithology.

Martin, F. W., Pospahala, R. S. & Nichols, J. P. (1979). Assessment and population monitoring of North American Migratory Birds. In *Environmental Monitoring, Assessment, Prediction and Management. Certain Case Studies and Related Quantitative Issues*, ed. J. Cairns, G. P. Patil & W. E. Waters, pp. 187–239. Fairland, International Cooperative Publishing House.

Moss, D. (1985). Some Statistical checks on the BTO Common Bird Census Index – 20 years on. In *Bird Census and Atlas Studies*, ed. K. Taylor, R. J. Fuller & P. C. Lack, pp. 175–9. Tring, British Trust for Ornithology.

O'Connor, R. J. (1981). The influence of observer and analyst efficiency in mapping method censuses. *Studies in Avian Biology*, **6**, 372–6.

O'Connor, R. J. (1985). Long-term monitoring of British bird populations. *Ornis Fennica*, **62**, 73–9.

O'Connor, R. J. & Hicks, R. K. (1980). The influence of weather conditions on the detection of birds during common birds census fieldwork. *Bird Study*, **27**, 137–51.

O'Connor R. J. & Shrubb, M. (1986). *Farming and Birds*, Cambridge, Cambridge University Press.

Peach, W., Baillie, S. & Underhill, L. Survival of British Sedge Warblers *Acrocephalus schoenobaenus* in relation to West African rainfall. *Ibis*, in press.

Reynolds, C. M. (1979). The heronries census: 1972–77 population changes and a review. *Bird Study*, **26**, 7–12.

Reynolds, R. T., Scott, J. M. & Nussbaum, R. A. (1980). A variable circular-plot method for estimating bird numbers. *Condor*, **82**, 309–13.

Root, T. (1988). *Atlas of Wintering North American Birds. An Analysis of Christmas Bird Count Data*. Chicago, The University of Chicago Press.

Rothery, P., Wanless, S. & Harris, M. P. (1988). Analysis of counts from monitoring Guillemots in Britain and Ireland. *Journal of Animal Ecology*, **57**, 1–19.

Sharrock, J. T. R. (1974). The changing status of breeding birds in Britain and Ireland. In *The Changing Flora and Fauna of Britain*, ed. D. L. Hawksworth, pp. 204–20. London, New York, Academic Press.

Sharrock, J. T. R. (1976). *The Atlas of Breeding Birds in Britain and Ireland*. Tring, British Trust for Ornithology.

Tapper, S. & Cook, S. (1988). Game bag records: the keystone of our research programme. *The Game Conservancy Review*, **1988**, 27–31.

Tatner, P. (1982). Factors influencing the distribution of Magpies *Pica pica* in an urban environment. *Bird Study*, **29**, 227–34.

Taylor, K., Fuller, R. J. & Lack, P. C. (1985). *Bird Census and Atlas Studies.* Proceedings of the VIII International Conference on Bird Census and Atlas Work. Tring, British Trust for Ornithology.

Titus, K., Fuller, M. R. & Ruos, J. L. (1989). Considerations for monitoring raptor population trends based on counts of migrants. In *Raptors in the Modern World*, ed. B.-U. Meyburg & R. D. Chancellor, pp. 19–32. Berlin, London, Paris, WWGBP.

Wild Bird Society of Japan and contributors (1982). *A Field Guide to the Birds of Japan*. Tokyo, Kodansha Int.

Winstanley, D., Spencer, R. & Williamson, K. (1974). Where have all the whitethroats gone? *Bird Study*, **21**, 1–14.

12
Freshwater biological monitoring

Introduction

IN 1989, the Secretary General of the International Council of Scientific Unions (ICSU) wrote the following:

water is the earth's most distinctive constituent

J. W. Maurits la Riviere (1989)

and then went on to describe how water is becoming an increasingly scarce resource while at the same time agriculture, industry and populations expand. Conservation of wildlife, protection of fish stocks and maintenance of amenity areas are other reasons for maintaining water quality. In Britain alone, the importance of water as a resource is demonstrated by the range of organizations which have an interest in the surveillance and monitoring of freshwater (Table 12.1). Industrialized countries throughout the world have different forms of legislation requiring certain standards of water quality and in most cases that legislation is enforced mainly to restrain levels of pollution and to ensure adequate public water supplies. Enforcement of legislation can, however, only take place if adequate monitoring has been undertaken and that monitoring requires appropriate funding, staffing and logistical support. Such funding and support is not always available. For example, Britain established The National Rivers Authority (NRA) in 1989 as a watchdog organization, but the effectiveness of that organization came under much criticism in the following year because lack of funds seems to have prevented adequate staffing and support. The result has been that many industrial and agricultural discharges into rivers have occurred in the absence of proper monitoring.

The ecological principles and the variables used for monitoring freshwater have already been described. In Chapters 2, 7 and 8, for example, we considered some of the diversity and biotic indices used for

Table 12.1. *Organizations responsible for surveillance and monitoring of freshwater in Britain*

Organization responsible	Surveillance/monitoring scheme	Dates and frequency of observations
University of Aston, Applied Hydrobiology	Benthic invertebrates of the River Cole	Annually since 1950
New College, London, Botany Dept.	Planktonic and other algae and zooplankton in Virginia Water	Weekly since 1958
University College of Wales, Cardiff, Botany Dept	Algae, bryophytes, macrophytes of certain rivers in South Wales, particularly the Usk	Since 1958, at varying intervals of time
Severn Trent Water Authority	Species lists for the Bristol Avon River Authority Area and Biological Assessment of Pollution	Irregular survey, 1935–71, 1950–75
South West Water Authority	Salmon in various Devon rivers	Since 1962, several censuses
Thames Water Authority	Plankton in Rivers Thames and Lee	Weekly or fortnightly since 1935
Welsh Water Authority	Salmon and Sea Trout and some other fish in South West Wales	Annually since 1952
Severn Trent Water Authority	Macro-invertebrates of the Trent	Bi-annually at *c.*600 sites since 1956
Severn Trent Water Authority	Freshwater fish in the Trent area	Irregularly since 1955
Wye River Authority (now Welsh National Development Water Authority)	Salmon counts on river Wye	Annually since 1903
Central Electricity Research Laboratory, Nottingham	Invertebrate communities in Lincolnshire	Species lists and numbers for 10 years (1960–69)
Field Studies Council	Brown Trout and Perch in Malham Tarn. Other taxa irregularly.	Angling returns for 25 years since 1947
Freshwater Biological Association River Laboratory	Fish in River Frome, East Stoke	Since 1964
FBA Windermere	Physical, chemical and biological data on the Cumbrian lakes	Since 1930 or earlier
Ministry of Agriculture, Fisheries and Food	Salmon and Sea Trout. Continuing census of ascending and descending fish on the River Axe, Devon	Since 1960
Department of Agriculture and Fisheries for Scotland	Salmon(1) sample counts on all ascending and descending fish and population estimates of young	Since 1966

Table 12.1. (*cont.*)

Organization responsible	Surveillance/monitoring scheme	Dates and frequency of observations
	fish in Girnock Burn, Aberdeenshire (2) sample counts of all ascending and descending fish in North Esk, Angus	Since 1962
	(3) sample counts of ascending and descending fish in River Meig, Ross-shire	Since 1957

Sources: Berry 1988, NERC 1976.

monitoring water quality. The aim of this chapter is to assess some case studies and the applications of indices for monitoring water quality, especially freshwater pollution.

Biological and chemical monitoring

Biological and chemical monitoring of pollution can and should supplement each other, though either could provide reasonable indications of the effects of pollutants such as organic effluents. Hynes illustrated the chemical and biological effects of organic pollution very clearly in 1960 with a now much published figure (Fig. 12.1). The use of biological organisms for monitoring water pollution has sometimes been criticized because it is a lengthy process (and therefore expensive), there is a lack of standardization of techniques, interpretation of biological data is complex and there is a lot of variation in the organism's response. Most of these and other disadvantages have recently been dismissed as myths about the barriers to biological monitoring techniques (Cullen 1990).

Chemical monitoring has had to become very sophisticated in recent years and this has been largely in response to the high standards enforced by some agencies such as the US Environment Protection Agency. Advances in computing and associated technology have now made it possible to detect very small concentrations of some pollutants within a very short period of time (Table 1.3). The examples given by Orio (1989) in that table, clearly demonstrate that chemical monitoring can be very sophisticated indeed.

However, chemical monitoring when used alone has its limitations and these limitations can be attributed to the following: difficulties associated with temporal and spatial sampling, potential synergistic effects, the possibility of bioaccumulation in ecosystems. That is, without continuous

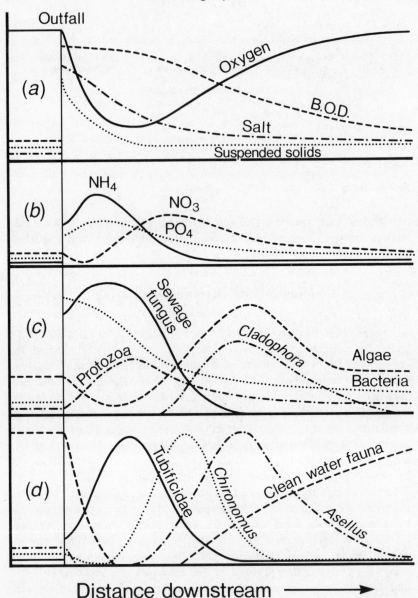

Fig. 12.1. Hypothetical effects of organic pollution in a river: *a* & *b*, physical and chemical changes, *c*, changes in micro-organisms, *d*, changes in larger organisms. Redrawn from Hynes (1960).

chemical monitoring there is a possibility that unusually high levels of pollutants (unpredictable events) could remain undetected but have serious effects on the aquatic ecosystem. In the absence of biological monitoring the combined effects of pollutants can not be assessed and chemical monitoring used alone may indicate deceptively low levels of pollutants which if accumulated could lower water quality below required standards and be harmful to aquatic life. Surprisingly, we know very little about the long-term effects of most chemical pollutants on aquatic ecosystems and biological processes. The impacts of radioactive wastes on freshwater and marine ecosystems especially pose very serious threats yet there is negligible monitoring of the influences of radiation on aquatic ecosystems.

The biochemical oxygen demand (BOD) is one common standard applied to monitoring and surveillance of freshwater. Often considered to be an aspect of chemical monitoring, it is in reality based on a biological process and should therefore be regarded as a basis for biological monitoring. The BOD is the ability of a given volume of water to use up oxygen over a period of five days at a temperature of 18 °C. Organic matter in the sample of water decomposes and the amount of oxygen thus consumed is then calculated. The range of BOD values and meaning of those values is compared with qualitative indices or biotic scores such as the Trent Biotic scores and Chandler Biotic scores in Table 12.2. Biotic scores are based on the presence or absence of certain taxa and the score is weighted according to the known tolerance of those taxa to pollution. Another common index, but a quantitative index, used to monitor water quality is the diversity index in its many forms (see p. 128).

Monitoring water quality

The following examples of monitoring freshwater have been selected, not because they are all particularly successful in monitoring water quality but because they offer a basis for discussion about the methods and the interpretation of the results especially with regard to pollution. Very useful and detailed aspects of the biology and ecology of water pollution can be found in Mason (1981).

The Water Quality Index

Determination of water quality requires value judgements no matter how many variables are measured and often there is conflict about which variables should be used. With the aim of promoting effective communication regarding the variables used for measuring water quality yet

Table 12.2. *Comparison of* BOD *with two biotic indices*

mg/l	General characteristics	Designation	Trent Index	Chandler Index
2	Salmonids Plecopterans Ephemeropterans Trichopterans Amphipods	Very clean	9–10	900 +
2–3	Coarse fish All above groups represented	Clean	7–10	500–900
2–3	Coarse fish Above groups restricted except Amphipods Trichopterans and Baetids	Clean	6–8	300–500
3–5	Few fish, Trichopterans restricted Baetids rare; *Asellus* dominant; Molluscs and leeches present	Fairly clean	5–6	110–400
5–10	As above, but no Baetids, Molluscs and leeches present	Doubtful	3–5	45–300
5–10	No fish, otherwise as above	Doubtful	2–4	15–80
10 +	No fish, Oligochaeta and red chrionomids only	Bad	1–3	9–20
10 +	None of above: only air breathing *Eristalis tenax*	Bad	0–1	0–10

From Hellawell 1978, Marstrand 1973.

retaining a simple index, a Water Quality Index (WQI) was developed in 1970 at the National Sanitation Foundation in the USA. The WQI can be expressed in the following form:

$$WQI = \sum_{i=1}^{n} w_i q_i \quad \text{(Brown } et\ al.\ 1972)$$

where WQI is a number between 0 and 100; q_i the quality of the ith variable, a number between 0 and 100; w_i the unit weight of the ith variable, a number between 0 and 1; and n is the number of variables.

After consultation with a panel of persons having expertise in water quality management, the following nine variables were selected for the index: dissolved oxygen, faecal coliforms, pH, BOD, nitrate, phosphate, temperature, turbidity, and total solids. Most of these variables are common to other schemes for assessing water quality. Because different experts apply different weighting to each variable, water quality was presented as a series of graphs, based on the respondent's views of what constituted low or high quality. This procedure, known as 'normalization of environmental indicators' (that is, converting measured indicators into a number or index) was described on p. 143 with reference to environmental indicators. With respect to the WQI, there was a close agreement for some parameters and poor agreement for others when normalization was undertaken (Fig. 12.2). The WQI has been advocated as being suitable for general usage as a uniform method reflecting the quality of water.

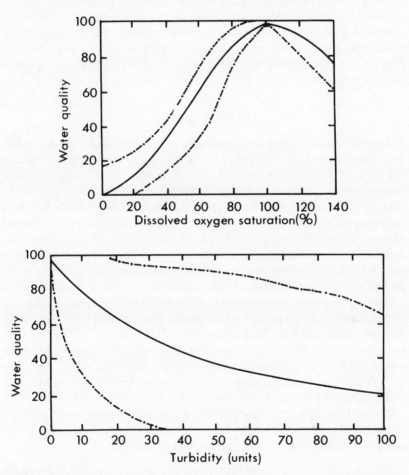

Fig. 12.2. Arithmetic means (solid lines) and 80% confidence limits (dashed lines) for two water quality parameters. Modified from Brown *et al.* 1972.

Indicators

For many years now, biological monitoring of freshwater has included the use of variables such as species composition, species richness, diversity, similarity, productivity, biomass and biotic indices based on community structure. Key or critical species such as commercially valuable fish or game fish have been used to assess pollution in aquatic systems and have been used in biological early warning systems (Cairns, Patil & Waters 1979, Gruber & Diamond 1988) and in the same manner, the concept of indicator species has been widely applied. More recently there have been some exciting developments combining the use of diatoms (as indicator species) and laser holography (see p. 99) for monitoring water quality. There is no general rule as regards choice of organisms for monitoring or surveillance; monitoring has to be based on those organisms which are most likely to provide the most appropriate information for the questions being asked (Mason 1981).

Toxicity testing

The use of aquatic organisms for assessment of the effects of pollutants and for continuous toxicity testing is not new and has been long practised in the USA in a very sophisticated manner (Cairns 1975, 1988, Cairns & Dickson 1978, Gruber & Diamond 1988). Toxicity tests have been used to provide data for several reasons (Buikema, Niederlehner & Cairns 1982) including the following: to assess effects on organisms exposed to sublethal concentrations of a toxic substance for part or all of a life cycle, to determine which pollutant is most toxic, to determine which organism is most sensitive, to assess changes in levels of toxicity, to determine if the pollutant is within regulatory standards and to determine effects of periodic changes in concentrations of the pollutant. Historically, toxicity tests have made much use of various species of adult fish (see p. 96), partly because more seems to be known about the ecology, physiology and behaviour of fish than other organisms. However, arguments against the use of fish in toxicity tests include the observation that many other taxa such as diatoms and macro-invertebrates are more sensitive to pollutants in water than are fish.

In the USA, the EPA has published criteria for determining which organisms should be used in toxicity testing and these and other suggested criteria include the following (Buikema *et al.* 1982).

1. The organism is representative of an ecologically important group (in terms of taxonomy, trophic level or niche).

2. The organism occupies a position within a food chain leading to many other important species.
3. The organism is widely available, is amenable to laboratory testing, easily maintained and genetically stable so uniform populations can be tested.
4. There are adequate background data on the organism (physiology, genetics, taxonomy, role in the natural environment, behaviour).
5. The response of the organism to the toxic substance should be consistent.
6. The organisms should not be prone to disease, excessive levels of parasitism or physical damage.

Many of the environmental standards in the USA have been based on the use of single-species toxicity testing and although these tests provide many useful data and although the response of one very sensitive species may reflect other responses at higher levels of biological organization, the tests have been questioned, especially in the light of recent advances in our knowledge of the complexities of ecosystems. Pontasch, Niederlehner & Cairns (1989) have argued, for example, that although single-species toxicity tests can efficiently examine relative toxicity, they may not be the most accurate method for predicting responses in receiving ecosystems because single-species tests do not take account of interactions between species and are sometimes conducted under physical and chemical conditions which are dissimilar to the natural environment. Assessments of single versus multispecies toxicity testing have been undertaken by Cairns and his colleagues; a recent consensus for the use of multispecies testing resulted in the support of 12 ideas which have been summarized in Table 12.3.

Cairns (1979) has long insisted that there is at present no reliable way, other than using an organism, to determine the integrated, collective impact of pollution on free-living aquatic organisms. The rationale underlying the use of toxicity testing as outlined above has formed the basis of proposals for in-plant and in-stream monitoring of water quality (Cairns, Dickson & Westlake 1977). At its simplest, this means that testing of the water, by biological monitoring, would take place between before the effluent discharge (in-plant) and at various locations in the stream below the effluent outfall. The costs of such sophisticated biological monitoring, especially automated biological monitoring using living organisms as sensors., often provide the reason for rejection of such an approach in favour of chemical monitoring or use of biotic indices (see question 3 below and also Chapter 8).

A useful, thought-provoking set of eight questions for an assessment of

Table 12.3. *Suggested ideas for assessing the effects of chemicals on ecosystems*

1. Multispecies tests can provide useful information for assessing the hazards of chemicals to aquatic and terrestrial organisms.
2. The end points (response parameters) measured in multispecies tests are not as decisive as the end points measured in single species tests.
3. Multispecies tests can be powerful analytical tools. The capability of isolating ecological functions for study (while retaining a certain degree of complexity) is the strongest attribute of multispecies test.
4. Multispecies tests can be used as a 'conceptual bridge' between more complex systems and less complex systems.
5. Multispecies testing has the ability to isolate structural and functional components of ecosystems which reduces variability.
6. Multispecies tests may be less sensitive to stress caused by a chemical when compared to the acute or chronic responses of uniform test organisms of a single species.
7. Multispecies tests are more complex than those conducted with single species and may exhibit more variability in structure and function.
8. The selection of the response parameter used to assess the effects of a chemical or effluent in a multispecies test is crucial because this selection can influence the interpretation of the results. Care must be taken to use state and rate variables correctly.
9. Because of the potential they offer in assessing the impact of chemicals on 'ecosystem level' functions, further research on multispecies toxicity tests is encouraged.

From Cairns (1986).

the conceptual soundness of in-plant and in-stream biological monitoring has been suggested by Cairns *et al.* (1979). In an abbreviated form these questions are as follows.

1. Will the system detect spills of lethal materials before they reach the outflow?
2. If only one organism is used for toxicity testing (e.g. a Bluegill Sunfish ...) will this species be more tolerant to some pollutants than to others?
3. Is it possible to monitor abiotic variables and achieve the same results at lower cost and with greater efficiency?
4. Since the biological monitoring can not identify the pollutant, is it possible to combine successfully both biological and chemical monitoring?
5. Will the behaviour of the organism be so reliable as not to cause 'false alarms' and possible expensive closing down of the plant?
6. Should an organism indigenous to each system be used or should

different species be used at different points in the in-plant and in-stream systems?

7. Is it possible to use in-plant biological monitoring systems to detect the presence of pollutants which are either acutely lethal or not lethal in the short-term but have long-term effects by way of bioaccumulation?

8. Are the in-plant biological monitoring systems only for a very large industry or is it possible to develop reliable, small in-plant biological monitoring that can be used by a person inexperienced in monitoring?

Monitoring effects of refinery aqueous effluents

British Petroleum (BP) have undertaken freshwater biological monitoring of Crymlyn Bog in South Wales over several years with the aim of looking at the effects on the fauna of refinery aqueous effluents and other souces of pollution (Girton 1983). A further objective was to assess different biological monitoring techniques for sampling and analysis of data. Hand-net sampling of the fauna took place at ten sampling sites on an annual basis and usually during mid-August. Three sampling sites were upstream of the effluent outfalls and seven were downstream. Sampling station No. 1 was subdivided into four sites, one above a quarry and three below a quarry outfall.

Sampling included both qualitative and quantitative methods. At each sampling station, three hand-net sweeps of 30 seconds duration using the methods of sampling recommended by the Department of the Environment (DOE 1978) were made of the benthos and surrounding macro-vegetation. Sediment and substratum material were washed through 710 micrometre mesh netting and the contents of the net preserved in 4 per cent formalin. Relative abundance was scored as follows: rare (1–2), occasional (3–10), common (11–50), abundant (51–100), very abundant (greater than 100).

Quantitative sampling was based on colonization samplers with a standard unit of substratum known as the Standard Aufwuchs Unit (S. Auf U. (Fig. 12.3)), and also cylinder and core sampling of the benthos (see Chapter 4 for comments on sampling methods and appropriate references). Water chemistry was also recorded but not described in this very condensed and selective account.

In order to aid the interpretation of the data in terms of water quality, the number of taxa, the Trent Biotic Index and the Chandler Score were calculated in a form presentable to management. Standardized sampling methods proved difficult because of changes in substrata and so therefore the Sorensen Coefficient of Similarity (see p. 132) was used to compare material obtained with different sampling methods. A diversity index based on a simplification of the Shannon–Wiener Index was employed.

The different sampling methods, not surprisingly, collected different numbers of taxonomic groups (Fig. 12.3). The analysis using Sorensen's Similarity Coefficient (Fig. 12.4) showed that similarity between colonization data and net data was highest and that there was poor similarity between net data and core data at stations 2,4 and 7. The results from the colonization sampling proved particularly successful in this kind of monitoring.

Overall, the decreasing number of taxa (Fig. 12.5) and the diminishing

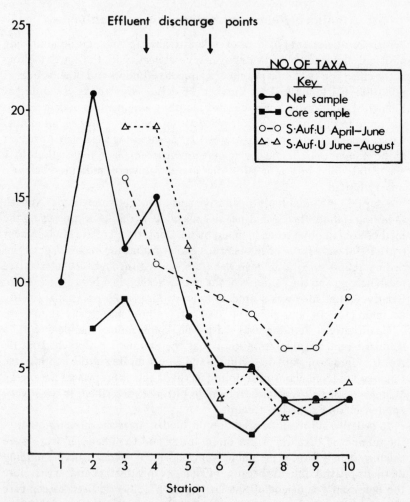

Fig. 12.3. Number of taxa collected by different sampling methods for the BP monitoring programme of Crymlyn Brook during 1982. From Girton (1983).

Effluent discharge points

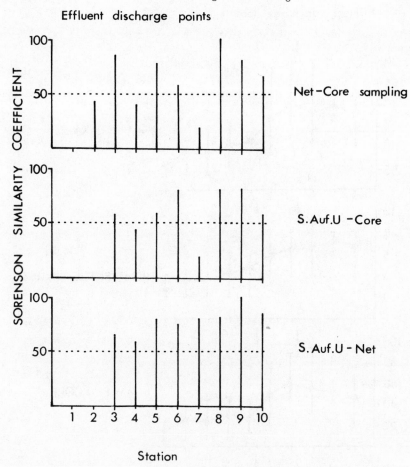

Net-Core sampling

S.Auf.U -Core

S.Auf.U - Net

Station

Fig. 12.4. Similarity indices of different sampling methods used for the BP monitoring programme of Crymlyn Brook during 1982. From Girton (1983).

scores for the three biotic indices at the ten sites was consistent. The first site was affected by ferruginous waste from a quarry which eliminated species unable to withstand low oxygen levels and/or colloidal particles. The increased values of the indices and increase in taxa at station 2 was a result of decreasing effect of the quarry run-off but also reflected organic enrichment from agricultural run-off. Downstream of station 5, the refinery effluent had eliminated most pollution sensitive species and the stream community was dominated largely by exploiter-type indicator groups such as Tubificidae, Chironomidae and *Lymnaea peregra*.

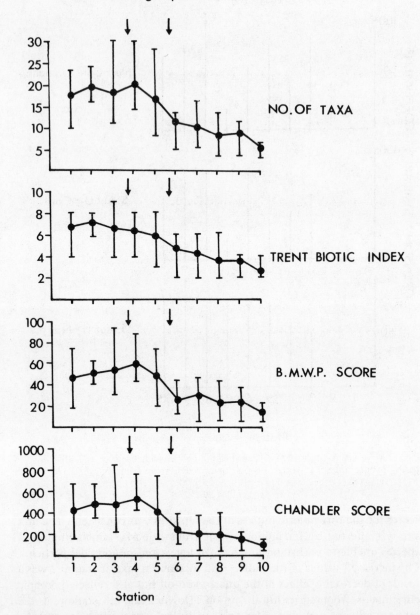

Fig. 12.5. Comparison of the number of taxa and biotic indices resulting from the BP monitoring programme of Crymlyn Brook, 1978–82. The effluent discharge points are shown by the arrows. From Girton (1983).

The BMWP biotic scores were similar in magnitude and resolution to the Chandler Scores but the former seemed to be more sensitive (there was larger variation) when applied to the core samples (Fig. 12.6). In general the conclusion was that the BMWP Score had proved extremely successful, especially as it relies on a relatively crude level of identification. All biotic indices were, however, useful because they facilitated rapid interpretation of data. A word of caution seems relevant here because the value of biotic scores and the relationship between biotic scores and other measures of water quality will vary from locality to locality. For example, Pinder & Farr (1987*a, b*) in their research on the surveillance of chalk streams found some interesting results when comparing biotic indices with BOD. They calculated both Chandler and BMWP scores, but also an average score per taxon was calculated for each biotic score by dividing the total score by the number of taxa contributing to it. They found that the Chandler Score did not show a significant correlation with any measure of water quality, although the average Chandler Score per taxon was significantly and negatively correlated with BOD and levels of dissolved organic carbon (DOC). Values of the BMWP Score were significantly negatively correlated only with DOC, whereas the average BMWP Score per taxon showed a highly significant negative correlation with BOD.

This five-year BP monitoring programme provides us with a generalized account but has proved to be an extremely useful basis for monitoring the responses of benthic populations to changes in levels of pollution. Standardized sampling at all stations had proved difficult and the amount of material produced by various sampling methods at 13 stations and substations was large, which in turn demanded large amounts of laboratory time for sorting, identifying and analysing. The report for this monitoring programme, not surprisingly, includes a recommendation for more selective sampling based on the colonization method.

Classification of rivers and lakes for monitoring and surveillance

The concept of community structure has long been used as a basis for monitoring and surveillance of water quality. For instance, in Pennsylvania, Patrick (1949) used histograms showing comparative abundance of different species to assess the degree of pollution and she pioneered subsequent uses of the community structure concept in river classification and river monitoring schemes.

More recently, Savage (1982) has shown that lakes in Britain can be classified on the basis of an analysis of affinities between communities of Water Boatmen (Corixidae). The method used by Savage and involving the

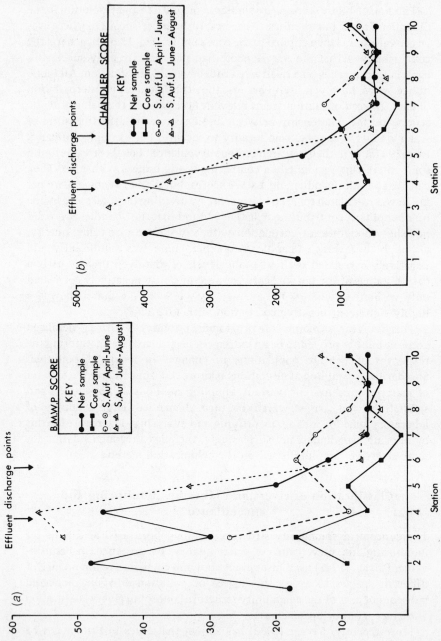

Fig. 12.6 The BMWP values (a) and the Chandler biotic indices (b) recorded for

determination of similarity indices followed by cluster analysis has been
described in Chapter 7. In brief, he showed a close relationship between
species of Corixidae (*Sigara* spp. and a *Callicorixa* sp.), conductivity and
organic load. Two species, *S. scotti* and *S. concinna*, are confined to waters
of low and high conductivity respectively. The species *S. falleni* and *C.
praeusta* tend to occur in areas of water of intermediate size and high
conductivity; the former being more numerous. The species *S. distincta*
occurs in ponds and lakes of relatively low conductivity but is replaced by
S. scotti at very low conductivity. The species *S. dorsalis* is found at all but
the lowest conductivity levels but tends to be replaced by *S. falleni* in
water bodies of high conductivity and intermediate size. On the basis of
these results, Savage suggested that the species could usefully be arranged
in three contrasting pairs:

scotti ...   *concinna* ...

distincta .... *praeusta*

dorsalis .... *falleni*

low conductivity − high conductivity

oligotrophic − eutrophic

Rather than basing a classification on either biotic or abiotic variables,
there has been a tendency to use combinations of variables. In Britain, for
example, a Department of the Environment and National Water Council
Biological Monitoring Working Party (BMWP) was given the brief of
developing a simple and efficient method for assessing biological quality of
rivers. A simple scoring system with a broad classification based on macro-
invertebrates was eventually recommended. The Biological Monitoring
Working Party score system (Table 8.1) has already been described in
Chapter 8. By way of contrast a Department of the Environment river
classification system was developed (HMSO 1985) with mainly a simple
chemical basis (Table 12.4). The DOE classification system is based on a
number of variables such as dissolved oxygen saturation levels, BOD and
levels of ammonia. The classifications are normally based on a 95 per cent
basis (that is parameters which are expected to be achieved with 95 per
cent of the samples taken).

Following the development of the Biological Monitoring Working Party
Score system, rapid appraisal of river quality was possible using
identification at family-level. Scientists at the Institute of Freshwater
Ecology and at the Freshwater Biological Association (FBA) have looked
closely at the use of 'families' of invertebrates for monitoring freshwater
communities and have now developed a river invertebrate prediction and
classification system which can be used to monitor water quality and

Table 12.4. *The DOE river classification*

River class	Quality criteria Class limiting criteria (95 percentile)	Remarks	Current potential uses
1A	(i) Dissolved oxygen saturation greater than 80%. (ii) Biochemical oxygen demand not greater than 3 mg/l. (iii) Ammonia not greater than 0.4 mg/l. (iv) Where the water is abstracted for drinking water, it complies with requirements for A2** water. (v) Non-toxic to fish in EIFAC terms (or best estimates of EIFAC figures not available).	(i) Average BOD probably not greater than 80%. (ii) Visible evidence of pollution should be absent.	(i) Water of high quality suitable for potable supply abstractions and for all other abstractions. (ii) Game or other high class fisheries. (iii) High amenity value.
1B	(i) DO greater than 60% saturation. (ii) BOD not greater than 5 mg/l. (iii) Ammonia not greater than 0.9 mg/l. (iv) Where water is abstracted for drinking water, it complies with the requirements for A2** water. (v) Non-toxic to fish in EIFAC terms (or best estimates if EIFAC figures not available).	(i) Average BOD probably not greater than 2 mg/l. (ii) Average ammonia probably not greater than 0.5 mg/l. (iii) Visible evidence of pollution should be absent. (iv) Waters of high quality which cannot be placed in Class 1A because of high proportion of high quality effluent present or because of the effect of physical factors such as canalization, low gradient or eutrophication. (v) Class 1A and Class 1B together are essentially the Class 1 of the River Pollution Survey.	Water of less high quality than Class 1A but usable for substantially the same purposes.
2	(i) DO greater than 40% saturation. (ii) BOD not greater than 9 mg/l. (iii) Where water is abstracted for	(i) Average BOD probably not greater than 5 mg/l. (ii) Similar to Class 2 of RPS.	(i) Waters suitable for potable supply after advanced treatment. (ii) Supporting reasonably good coarse

requirements for A3** water. (iv) Non-toxic to fish in EIFAC terms (or best estimates if EIFAC figures not available).	pollution other than humic colouration and a little foaming below weirs.	(iii) Moderate amenity value.
3 (i) DO greater than 10% saturation. (ii) Not likely to be anaerobic. (iii) BOD not greater than 17 mg/l.	Similar to Class 3 of RPS.	Waters which are polluted to an extent that fish are absent or only sporadically present. May be used for low grade industrial abstraction purposes. Considerable potential for further use if cleaned up.
4 Waters which are inferior to Class 3 in terms of dissolved oxygen and likely to be anaerobic at times.	Similar to Class 4 of RPS.	Waters which are grossly polluted and are likely to cause nuisance.
X DO greater than 10% saturation.		Insignificant watercourses and ditches not usable, where objective is simply to prevent nuisance developing.

Notes: (a) Under extreme weather conditions (e.g. flood, drought, freeze-up), or when dominated by plant growth, or by aquatic plant decay, rivers usually in Classes 1, 2 and 3 may have BODs and dissolved oxygen levels, or ammonia content outside the stated levels for those Classes. When this occurs the cause should be stated along with analytical results.

(b) The BOD determinations refer to 5 day carbonaceous BOD (ATU) Ammonia figures are expressed as NH_4.

(c) In most instances the chemical classification given above will be suitable. However the basis of the classification is restricted to a finite number of chemical determinands and there may be a few cases where the presence of a chemical substance other than those used in the classification markedly reduces the quality of the water. In such cases, the quality classification of the water should be downgraded on the basis of the biota actually present, and the reasons stated.

(d) EIFAC (European Inland Fisheries Advisory Commission) limits should be expressed as 95% percentile limits.

*This may not apply if there is a high degree of re-aeration.

**EEC category A2 and A3 requirements are those specified in the EEC Council Directive of 16 June 1975 concerning the Quality of Surface Water intended for Abstraction of Drinking Water in the Member States.

From HMSO (1985). Reproduced with the permission of the Controller of Her Majesty's Stationery Office.

pollution. This system is aptly called 'RIVPACS' and is based on the assumption that the presence of certain taxonomic groups of invertebrates in rivers will depend on levels of certain physical and chemical variables. The programme to establish RIVPACS commenced at the FBA in 1977 with two main objectives: to develop a classification system for unpolluted running-water sites in Great Britain based on macro-invertebrate fauna,

```
PROGRAMME RIVERS/:

RIVER: Moors                                    SITE: King's Farm

ENVIRONMENTAL DATA:

Water Width (m)                      3.93
Water Depth (cm)                    48.8
Boulders & Cobbles (%)               0.0

Pebbles & gravel     (%)             9.33
Sand                 (%)            16.33
Silt and clay        (%)            74.33

Mean substratum      (phi)           5.97
Altitude (m)                        19.0
Distance from source (km)           16.8
Slope   (m/km)                       1.44
Air temp. range (c)                 12.3
Mean air temp. (c)                  10.56
Total oxidized N (ppm)               4.26
Alkalinity (ppm CaCo3)             161.67
Chloride (ppm)                      12.53

PREDICTED TAXA (%), IN DECREASING ORDER OF PROBABILITY IN SAMPLE

* 99.9 Oligochaeta            * 83.6 Planorbidae
* 99.9 Chironomidae           * 81.4 Dysticidae
* 99.7 Baetidae               * 81.1 Planariidae
* 98.9 Sphaeriidae            * 79.8 Haliplidae
* 98.7 Glossiphoniidae        * 79.6 Tipulidae
* 98.6 Elminthidae            * 79.6 Lymnaeidae
* 98.3 Gammaridae             * 76.9 Polycentropodidae
* 93.4 Erpobdellidae          * 70.9 Ancylidae
* 93.2 Asellidae                59.4 Corixidae
* 93.1 Simuliidae             * 59.3 Piscicolidae
* 93.0 Hydrobiidae            * 59.3 Physidae
* 90.2 Limnephilidae          * 57.8 Rhyacophilidae
* 87.5 Hydropsychidae         * 55.7 Valvatidae
* 85.2 Caenidae               * 55.5 Gyrinidae
* 85.1 Hydroptilidae            55.5 Sialidae
* 84.7 Ephemerellidae           53.3 Psychomyiidae
* 83.9 Leptoceridae           * 52.6 Goeridae

end ..........
```

Fig. 12.7. A printout showing predictions of BMWP families to the 50% probability level based on 11 physical and chemical features. The asterisks indicate the families found after the standard sampling effort over three seasons. From Wright *et al.* 1988.

and to develop procedures for prediction of fauna which would be expected on the basis of environmental variables (Furse *et al.* 1984, Wright *et al.* 1989). The TWINSPAN programme for analysis of multivariate data was used to establish classification of 30 site groups of fauna based on 370 sites from 61 river systems. A manual for RIVPACS (Furse *et al.* 1986) provides details of the field and laboratory methods for taking the biological samples.

This technique provides equations for prediction of the probability of groups at new sites having sets of known environmental variables (Fig. 12.7). Any one of four sets of environmental variables may be used to predict the fauna at the following taxonomic levels: species, all families (including log categories of abundance) and restricted listings of BMWP families. An example for a river site in southern England, described by Wright *et al.* (1988, 1989), gives the probability of capture of each Family (in this instance to the 50 per cent level of probability). The observed families are indicated in Fig. 12.7. A comparison of observed and predicted numbers of BMWP families is then used as a basis for assessment of environmental stress affecting river communities.

References

Berry, R. J. (1988). *Biological Survey: Need & Network*. Report of a working party set up by the Linnean Society of London. London, PNL Press.

Brown, R. M., McClelland, N. I., Deininger, R. A. & O'Connor, M. F. (1972). A water quality index – crashing the psychological barrier. In *Indicators of Environmental Quality*, ed. W. A. Thomas, pp. 173–82. New York, London, Plenum Press.

Buikema, A. L., Niederlehner, B. R. & Cairns, J. (1982). Biological Monitoring part IV – toxicity testing. *Water Research*, 16, 239–62.

Cairns, J. (1975). Quality control systems. In *River Ecology*, ed. B. A. Whitton, pp. 588–612. Oxford, Blackwell Scientific Publications.

Cairns, J. (1979). Biological Monitoring – concept and scope. In *Environmental Biomonitoring, Assessment, Prediction, and Management – Certain Case Studies and Related Quantitative Issues*, J. Cairns, G. P. Patil & W. E. Waters, pp. 3–20. Fairland, USA, International Co-operative Publishing House.

Cairns, J. (1986). Multispecies toxicity testing: a new information base for hazard evaluation. *Current Practices in Environmental Science and Engineering*, 2, (E-3), 37–49.

Cairns, J. (1988). Politics, economics, science – going beyond disciplinary boundaries to protect aquatic ecosystems. In *Toxic Contaminants and Ecosystem Health; a Great Lakes Focus*, ed. M. S. Evans, pp. 1–16, John Wiley & Son.

Cairns, J. & Dickson, K. L. (1978). Field and laboratory protocols for evaluating the effects of chemical substances on aquatic life. *Journal of Testing and Evaluation*, 6, 81–90.

236 Monitoring in practice

Cairns, J., Dickson, K. L. & Westlake, G. F. (1977). *Biological Monitoring of Water and Effluent Quality.* Philadelphia, American Society for Testing and Materials.

Cairns, J., Patil, G. P. & Waters, W. E. (1979). *Environmental Biomonitoring, Assessment, Prediction, and Management – Certain Case Studies and Related Quantitative Issues.* Fairland, USA, International Co-operative Publishing House.

Cullen, P. (1990). Biomonitoring and environmental management. *Environmental Monitoring and Assessment,* 14, 107–14.

DOE (1978). *Methods of Biological Sampling. Handnet Sampling of Aquatic Benthic Macroinvertebrates.* London, HMSO.

Furse, M. T., Moss, D., Wright, J. J. & Armitage, P. D. (1984). The influence of seasonal and taxonomic factors on the ordination and classification of running-water sites in Great Britain and on the prediction of their macroinvertebrate communities. *Freshwater Biology,* 14, 257–80.

Furse, M. T., Moss, D., Wright, J. F., Armitage, P. D. & Gunn, R. J. M. (1986). *A Practical Manual for the Classification and Prediction of Macroinvertebrate Communities in Running Water in Great Britain.* Preliminary Version, East Stoke, Wareham, FBA River Lab.

Girton, C. (1983). *Freshwater Biological Monitoring of the Crymlyn Bog at BP Oil Llandarcy Refinery Limited, Neath, West Glamorgan, 1978–1982.* London, BP Environmental Control Centre.

Gruber, D. S. & Diamond, J. M. (1988). *Automated Biomonitoring: Living Sensors as Environmetnal Monitors.* Chichester, Ellis Horwood.

Hellawell, J. M. (1978). *Biological Surveillance of Rivers.* Stevenage, NERC, WRC.

HMSO (1985). *Methods of Biological Sampling. A Colonization Sampler for Collecting Macro-invertebrate Indicators of Water Quality in Lowland Rivers.* London, HMSO.

Hynes, H. B. N. (1960). *The Biology of Polluted Waters.* Liverpool University Press.

Marstrand, P. K. (1973). Using biotic indices as a criterion of in-river water quality. *Yb. Ass. River Auth.,* 1973, 182–8.

Mason, C. F. (1981). *Biology of Freshwater Pollution.* London, New York, Longman.

NERC (1976). *Biological Surveillance.* Working Party Reports. NERC. Publ. ser. B. p. 18.

Orio, A. A. (1989). Modern chemical technologies for assessment and solution of environmental problems. In *Changing the Global Environment. Perspective on Human Involvement,* ed. D. B. Botkin, M. F. Caswell, J. E. Estes & A. A. Orio, pp. 169–84. London, Academic Press.

Patrick, R. (1949). A proposed biological measure of stream condition based on a survey of the Conestaga Basin, Lancaster County, Pennsylvania, *Proceedings of the Academy of Natural Sciences, Philadelphia,* 101, 377–81.

Pinder, L. C. V. & Farr, I. S. (1978a). Biological surveillance of water quality – 2. Temporal and spatial variation in the macroinvertebrate fauna of the River Frome, a Dorset Chalk Stream. *Archives of Hydrobiology,* 109, 321–31.

Pinder, L. C. V. & Farr, I. S. (1978b). Biological surveillance of water quality – 3. The influence of organic enrichment on the macroinvertebrate fauna of a small chalk stream. *Archives of Hydrobiology,* 109, 619–37.

Pontasch, K., Niederlehner, B. R. & Cairns, J. (1989). Comparisons of single-species, microcosm and field responses to a complex effluent. *Environmental Toxicology and Chemistry,* 8, 521–32.

La Riviere, J. W. M. (1989). Threats to the world's water. *Scientific American*, **261**, 48–55.

Savage, A. A. (1982). Use of water boatmen (Corixidae) in the classification of lakes. *Biological Conservation*, **23**, 55–70.

Wright, J. F., Armitage, M. T., Furse, M. T. & Moss, D. (1988). A new approach to the biological surveillance of river quality using macroinvertebrates. *Verh. Internat. Verein. Limnol.*, **23**, 1548–52.

Wright, J. F., Armitage, M. T., Furse, M. T. & Moss, D. (1989). Prediction of invertebrate communities using stream measurements. *Regulated Rivers: Research and Management*, **4**, 147–55.

13

Insularization and nature conservation

Introduction

BIOTIC COMMUNITIES throughout the world have progressively become reduced in area as a direct result of human exploitation. During this process of reduction, the wildlife habitats and natural communities have become progressively more and more fragmented, resulting in remnants of habitats and communities being surrounded by different land uses. As the area of the natural communities becomes smaller and the number of fragments increases, there is an increasing incidence in the number of isolated plant and animal populations. This world-wide phenomenon of reduction, fragmentation and isolation of natural communities (insularization) has occurred over different time-scales and in different geografical scales in all parts of the world except Antarctica. Insularization of temperate forests for example has occurred over many hundreds of years whereas insularization of tropical forests has occurred only more recently on a large scale. For example current losses of tropical rainforests are about 76 000 sq km per annum (about one-third the area of Britain) and in some areas this rate of loss continues to rise (Figs 13.1 and 13.2).

One response to these losses has been to establish various kinds of protected areas, the objective being to conserve the remaining fragments. This is not to say that all protected areas have been created in response to insularization; many nature reserves have been designated as part of a conservation effort directed at a particular species or group of species. In some countries, the protected areas (such as national nature reserves) have been selected so as to represent the major ecosystems of that country. In other words, protected areas have been established throughout the world for wide-ranging objectives (Table 13.1) but two common and broad objectives have been the conservation of the landscapes and conservation

238

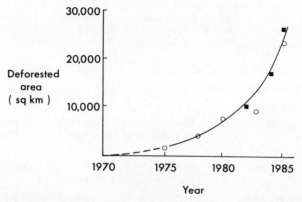

Fig. 13.1. Rate of loss of forests in Rhondonia (southeastern Amazon basin of Brazil). ■, data from satellites, ○, data from ground surveys. From Spellerberg & Hardes (1991) after Malingreau & Tucker (1988).

As deforestation progresses, it reduces the quality of life of millions of people in developing countries; their survival is threatened by the loss of the vegetation upon which they depend for their sources of household energy and many other goods. If tropical forests continue to be cleared at the current rate, at least 556 million acres (225 million hectares) will be cleared by the year 2000; if destruction of the tropical rain forests continues unabated, an estimated 10 to 20 per cent of the Earth's plant and animal life will be gone by the year 2000.

Reversing deforestation depends on political leadership and appropriate policy changes by developing-country governments to support community-level initiatives. The key ingredient is active participation by the millions of small farmers and landless people who daily use forests and trees to meet their needs.

J. Gustave Speth
President, World Resources Institute
WCED Public Hearing, Sao Paulo, 28-29 Oct 1985

Fig. 13.2. An extract from the World Commission of Environment and Development report, *Our Common Future* (WCED 1987). Reproduced with kind permission of Oxford University Press.

of wildlife. The degree of protection afforded by these protected areas is as varied as is the legislation of authority responsible for the particular protected area. Equally varied is the amount of available information about the species in the protected areas and certainly little monitoring of wildlife takes place even in the larger protected areas such as national parks. In Africa, for example, and almost certainly elsewhere the lack of monitoring

Table 13.1. *Protected areas in Britain: nature conservation and landscape conservation*

Designation	Statutory-basis/control
A. Nature Conservation Areas	
Biosphere Reserves	Countryside Commission. NCC.
Local Nature reserves (LNR)	Local planning authority in consultation with NCC (National Parks and Access to Countryside Act 1949).
Marine Nature Reserves (MNR)	Declared by the Secretary of State (Wildlife & Countryside Act 1981).
National Nature Reserves (NNR)	Selected, designated and managed by NCC (1949 Act).
NGO Nature reserves	Selected, designated and managed by non-governmental organization: RSPB, County Naturalist's Trusts.
Private Nature Reserves	Non-statutory, developed by private individuals
'RAMSAR' Sites[a]	Wetland Sites designated by Governments (Ramsar Convention on wetlands of international importance). NCC for Government.
Sites of Special Scientific Interest (SSSI)	Notified by NCC (1949 and 1981 Acts).
Special Protection Areas for birds	Member States of the EEC (Article 4 of the Council Directive on the Conservation of Wild Birds).
Biogenetic Reserves	Resolution (76) 17, Council of Europe, Committee Ministers. Not governed by a convention or directive.
B. Protected areas which incorporate nature conservation	
World Heritage Sites	The 'World Heritage Convention', UNESCO, 1975.
Areas of Outstanding Natural Beauty (AONB)	Countryside Commission (1949 Act).
Country Parks	Non-statutory but designated by Countryside Commission in conjunction with local authorities and private bodies.
Environmentally sensitive Areas	Designated by Ministry of Agriculture, Food and Fisheries. (Agriculture Act 1986.)
Green Belts	Defined by local authorities within structure plans.
Heritage Coasts	Non-statutory and designated jointly by local authority and Countryside Commission.

Table 13.1. (*cont.*)

Designation	Statutory-basis/control
MOD land	Some Ministry of Defence areas by virtue of their restricted access, have become and are now managed as wildlife reserves.
National Parks	Delineated by the Countryside Commission and declared under 1949 Act.
National Scenic Areas	Scottish equivalent of NPS and AONBS.
National Trust Land	National Trust Act of 1907.
Other areas or special areas	1. Some areas (The Norfolk Broads, New Forest, Somerset Wetlands) have strict planning controls. 2. Roadside verges, recreational land, allotments, and many other 'green areas' are sometimes managed as nature reserves.

[a]See Lyster (1985) for details about international wildlife law and conventions. *From* Spellerberg & Hardes (1991).

in national parks can be attributed to a basic problem of turnover in staff and the problems that that brings with reference to continuity and uniformity of methods. However, there are some notable national park wildlife monitoring programmes in Africa, one having been initiated in the Kruger National Park in 1975 (Huntley 1988). The current monitoring programme is based on fixed-point photographic stations plus aerial counts of 12 species of large mammals together with the collection of climate, fire and vegetation data.

Two interesting aspects of monitoring emerge from this problem of insularization and the establishment of nature reserves. One aspect concerns the need to monitor gaps in terms of the representativeness of the protected areas. That is, on the basis of surveys of ecosystems, national nature reserves are selected but this approach is flawed if the surveys of ecosystems have not been comprehensive. In addition to this, changes in land use will result in varying losses of different ecosystems and therefore the nature reserves are bound not to remain representative. To overcome this problem, there is a need to monitor the gaps in the ecosystems which are not conserved; something which is easier said than done. One of the greatest difficulties has been the lack of a reliable and comprehensive method for surveying, recording and storing data on ecosystems (see Chapter 14).

Several countries have begun to tackle the difficulties of surveying ecosystems (South Africa, Australia and Britain included) but there are some countries such as the United States where apart from some preliminary studies, no countrywide analysis of ecosystems exists (Burley 1988). Surprisingly one of the reasons for this lack of countrywide ecosystem information has been the lack of standardization for recording; we come back to this fundamental point in Chapter 14.

Monitoring ecosystems and monitoring gaps in the representativeness of protected areas are not discussed here. Rather, we look at an equally fundamental aspect of monitoring and that is the rate, extent and effect of insularization. Monitoring the rate and extent of insularization has not yet been undertaken anywhere in the world in a planned systematic and scientific fashion. However, there have been some attempts to monitor insularization in retrospect. There have also been some attempts to monitor changes in land use and this is discussed later in Chapter 14. Much attention has been directed towards the recording of losses (but rarely gains) of natural areas, habitats and communities in many countries throughout the world. Although it can be argued very strongly that there is a need to monitor the rate at which biotic communities are reduced and the extent to which they are fragmented, there is perhaps a greater need to monitor the levels of success or failure arising from attempts to reduce the impact of insularization. Equally important is the need to monitor the many effects of insularization on wildlife.

In this chapter we look first at examples of insularization and the response in terms of the establishment of protected areas, especially areas set aside for nature conservation. Secondly, the effects of insularization are described and these are then followed by examples of monitoring remnant populations and examples of monitoring the effects of insularization. The main aim of this chapter is to draw attention to a large gap in biological and ecological monitoring.

Losses of wildlife and rates of insularization

The rate and extent of insularization have been most severe in temperate regions and the main causes of losses in natural areas and subsequent insularization have been the extensive development of agriculture and forestry. Temperate closed forests have seemingly suffered the greatest cumulative losses, 32–35 per cent, followed by tropical woody savannahs and deciduous forests and tropical climax forests (Repetto & Gillis 1988). In Australia, the temperate forests have been decimated to such an extent that only small scattered remnants of eucalyptus woodlands remain in the man-made landscapes. To add to that folly, watersheds have been damaged or totally destroyed. In New Zealand, temperate indigenous forests have been cleared at the rate of 1250 ha per year to provide woodchip for the paper industry. The resultant effect on New Zealand wildlife has been estimated (not studied in detail) to be severe. In Europe, the forests have been cleared throughout successive centuries, beginning in prehistoric times. Although some tracts of forests still remain and although there have been some gains in afforestation, the net result has been an ever increasing insularization of forests and woodlands (Darby 1956).

In Britain, reduction and fragmentation of forests commenced some 5000 years ago and now virtually all forests have gone from Britain's landscape. Only about 4 per cent cover of native woodland remains and only 1.5 per cent is both ancient and semi-natural. These woodlands that do remain are small, scattered isolated remnants and most are less than 20 ha in area (Peterken, 1991).

There has been much documentation of losses of natural and semi-natural vegetation in Britain since the early 1940s (see for example Tables 14.4, 14.5) and more recently organizations such as the Nature Conservancy Council (NCC) and Royal Society for Nature Conservation (RSNC 1989) have drawn attention to the serious and continual erosion of natural and semi-natural areas. As much as 30–50 per cent of semi-natural broadleafed woodland has gone since the 1940s. Species-rich lowland grasslands on chalk and limestone have been reduced to about 20 per cent and about 3 per cent of lowland grassland remains with any scientific interest. By way of comparison and on a smaller scale, insularization of the lowland heathlands of Britain can be traced in a more detailed manner, particularly with regard to the increase in number of heathland fragments (Fig. 13.3). A programme for monitoring the insularization of lowland heathlands has now been developed at the ITE Furzebrook Research Station (Webb 1990).

Perhaps the most dramatic example of extent in deforestation has been that during the 17th, 18th and 19th centuries in North America. As an example of the dramatic decline in area of woodlands, Whitney & Somerlot (1985) give figures for one representative county in the Midwest, showing that only about 85 to 90 per cent of the woodland had been cleared by the early 1900s.

The most recent examples of insularization come from tropical regions where forests have been devastated in an alarming manner (Fig. 13.2). For example, less than 6.8 per cent of the total area of the Atlantic forest of Minas Gerais in Brazil has any form of forest cover (Fonseca 1985). Recent monitoring via satellites has dramatically shown the vast areas of desolate, parched land where the forests have been destroyed. Losses of forests in and around Central America have received much publicity but the implications there are equally, if not as, severe as on ancient islands such as Madagascar, thought to have been one of the biologically richest areas in the world. One study using satellite and aerial photographs of Madagascar revealed that between 1950 and 1985, half the tropical rain forest had gone (Green & Sussman 1990).

With the exception of the very recent satellite monitoring of tropical South American forests (an essentially unused database according to Green & Sussman 1990), there has not been systematic surveillance or monitoring of the above examples of insularization. It is possible, however, to trace the rate and extent of insularization in a crude form by using historical data. In the South Island of New Zealand, for example, the losses and fragmentation of tussock grasslands can be documented from historical records (Fig. 13.4) but the limitations in terms of accuracy can also be seen in the maps shown in Fig. 13.4. Obviously the extent and distribution of tussock grasslands was mapped with less accuracy in pre-European times compared to 1840 and 1987.

Nature reserves

Throughout the world, some protected areas have been established with the objective of conserving and protecting landscapes, ecosystems and a wide range of endangered taxa. It is estimated that about four per cent of the world's land surface has been set aside for protection of biological

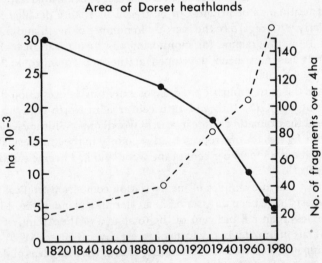

Fig. 13.3. The decline and fragmentation of lowland heathlands in southern England. From Spellerberg (1981).

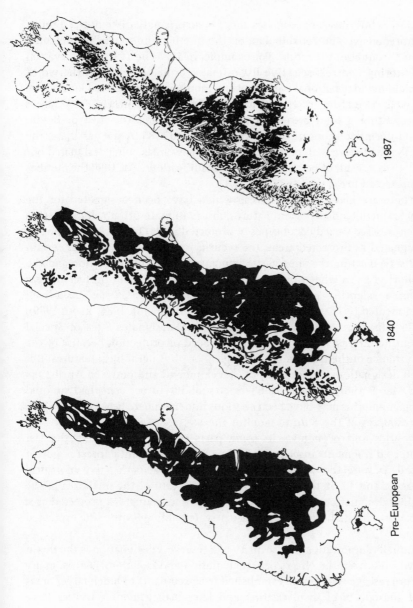

Pre-European   1840   1987

Fig. 13.4. The distribution of tussock grassland in New Zealand: an attempt at historical monitoring of insularization. Information kindly supplied by the Tussock Grasslands & Mountain Lands Institute, New Zealand.

resources but these are estimates and it is surprisingly difficult to establish comprehensive and reliable data on the number and extent of protected areas throughout the world. For example, in 1987 the IUCN Conservation Monitoring Centre's selective list of major protected areas in the world (which includes national parks, natural monuments, protected landscapes as well as nature reserves) had information on nearly 3000 areas representing a total area equivalent to about half the area of Brazil. That information concerned data collected up to 1985. A few years later the number had increased, partly as a result of new areas being designated but largely as a result of better data and communications. For 1990 the number of protected areas was 4679 (Fig. 2.1).

There are many kinds of areas which have been designated for the protection of landscapes and nature conservation. In Britain alone, there is an impressive varied list of types of protected areas (Table 13.1). Although designated as protected areas, the security of these areas is never certain and safeguarding the sites varies immensely with the legislation and the authority. Even what would seem to be highly important protected areas (from a scientific point of view) are not secure, as has been well documented by many organizations (e.g. Wildlife Link 1984, RSNC 1989). In Britain, the Nature Conservancy Council designates Sites of Special Scientific Interest (SSSIs) which are in need of protection because of the importance of the flora, fauna, geological or physiographical features. But such designations do not prevent 'development' and between April 1984 and March 1988, as many as 687 SSSIs (about 14 per cent) had suffered damage and this may have been a serious underestimate because of the lack of resources of the NCC to monitor the SSSIs.

Politics and economics in some parts of the world make the last protected fragments insecure, as is the case in the Atlantic forest region of Brazil. Despite the establishment of five national parks, five biological reserves and 17 State parks, (about 0.82 per cent of the total area of the region)·designation as a park does not assure its status as a protected area because of limited funds to purchase private holdings within the protected areas.

Equally important as the security of a reserve's designation is the threat to wildlife from the effects of fragmentation and isolation. That is, many nature reserves have been established from remnants of much larger areas and the effects of insularization and later management activities have important implications for the wildlife on the reserve.

Effects of insularization

The effects of insularization can be considered under two headings: firstly the immediate effects on wildlife caused by the process of fragmentation and isolation, and secondly the long-term effects caused by isolation and disturbances or perturbations (such as fire and trampling) on the remnant or insular community (see Verner, Morrison & Ralph 1986 for an extensive review of prediction and modelling of effects of habitat fragmentation). Reduction in area of natural communities affects population size and fragmentation affects dispersion, all of which leads to extinctions. Loss of habitat heterogeneity is also a product of insularization and those species with patchy distributions or those which utilize a range of microhabitats are especially vulnerable to losses in mosaics of habitats (Wilcove, McLelland & Dobson 1986). Other long-term consequences of insulariz-ation include the increase in number of invasive species, increases in levels of heavy metals and increases in levels of nutrients.

The insularization of tropical forests has resulted in species and population extinctions. Dramatic examples come from the tropical regions, and especially the 'Minimum Critical Size Ecosystem Projects' (Lovejoy et al. 1986). These projects commenced in 1979 and continue as the 'Biological Dynamics of Forest Fragment Project' (BDFF). Lovejoy and his colleagues have had access to newly created fragments of tropical forest ranging in area from 1 ha to 10 000 ha; although some species are favoured by newly created edges, Lovejoy and his colleagues have, in general, reported increased tree mortality, and diminishing numbers of species, as well as extinction of both plants and animals.

The effects of fragmentation and isolation on animal populations have been particularly well researched. Butcher et al. (1981) for example reported the effects of changes in vegetation and nearby urbanization on breeding bird species in Connecticut, and found that turnover rates (repeated extinctions and immigration of various taxanomic groups) of forest breeding birds were due to increasing isolation from similar forest habitat which later resulted in local extinctions of forest interior species. The investigation by Berstein et al. (1976) on the differential impact of forest degredation upon five species of monkeys in northern Colombia showed marked contrasts in species present, relative numbers and population densities; for example, one species Lagothrix, was virtually absent in remnant forests while other species such as Cebus, Alouatta and Saginus persisted there.

Insularization has both ecological and genetic implications and reduced variability has long been recognized as a feature of small and isolated populations (see Brakefield 1991 for a review of genetic implications of insularization). For example, the remaining forest patches of the Brazilian Atlantic forest may contain populations of many species which are too small to ensure that genetic variability is maintained (Fonseca 1985). In general there are considered to be three important processes which are causes of reduced variability in small, isolated populations; that is inbreeding, genetic drift and the 'bottleneck' and 'founder' effects (Miller 1979). Inbreeding increases the proportion of homozygotes to hetero-zygotes and therefore increases the incidence of recessive variants. Genetic drift is the change in gene frequency which results, in each generation, from the random sampling of alleles in panmictic populations. The effects of genetic drift are particularly noticeable in very small genetically isolated populations (see p. 303). The 'bottleneck' and 'founder' effects occur in very reduced populations or where a few individuals establish a population in a new area and in both cases few individuals are passing on a small fraction of the genetic material of the original population.

Nature reserves and monitoring insular communities

Most nature reserves are remnants of what were once much larger biotic communities and these remnant communities will experience ecological processes which occur as a result of insularization. For example and despite the establishment of large protected areas such as national parks, species losses and changes in species composition may take place. Newmark (1987), in describing his research on mammals in North American national parks, attributed post-establishment losses of mammals to the losses of habitat and the active elimination of fauna on land adjoining the parks. Biological invasions pose an especially difficult problem for the manage-ment of protected areas (see Usher 1988 for a review) and clearly fauna and flora of the surrounds will affect the ecology of the protected area. An example of how the surrounds can affect protected areas of fragmented heathland (a plant species poor community) in southern England has been well demonstrated by the work of Webb, Clarke & Nicholas (1984). They found that where structurally more diverse vegetation surrounded a small remnant heathland, there was a tendency for the invertebrate diversity to be greater; where structurally less diverse vegetation surround the heathland, few changes in invertebrate species diversity were detected.

Monitoring the effects of insularization

Monitoring the rate and extent of insularization of biotic communities is as important as monitoring the extent of any physical resource such as minerals or water but two other aspects of insularization would seem to be more important candidates for monitoring. It is, for example, more important to monitor the attempts to alleviate the effects of insularization such as the establishment of buffers zones and corridors. It is equally important to monitor populations in remnant communities and monitor the ecological and genetic effects of insularization. Such monitoring programmes could provide basic data for the formulation of management strategies for many kinds of protected areas including national parks and national nature reserves.

Monitoring the attempts to alleviate the effects of insularization

Ecological buffer zones and ecological corridors are two concepts which have been put forward as ways of diminishing any isolation effects of nature reserves. Buffer zones have been incorporated around reserves to reduce the effects of the surrounds and to conserve the 'ecological integrity' of the reserve. Ecological corridors (such as hedgerows, waterways and forest rides) have long been advocated as being important in aiding the dispersal of wildlife between fragmented habitats and the concept of corridors has been extended to networks (combination of reserves and corridors on a large scale (for a review see Spellerberg 1991). In certain regions of the world, for example in sub-Saharan Africa, reserves and buffer zones are on such a large scale that any monitoring of the buffer zones would require the application of satellite technology. This has been suggested by Vujakovic (1987) who has undertaken research on the large scale role of buffer zones and wildlife corridors around wildlife reserves in parts of Botswana. Mapping variation in canopy cover of low tree and shrub savanna, enabled Vujakovic (1987) to develop improvements in the classification of LANDSAT data which in turn provides a scale of imagery suitable for monitoring environmental changes and the effectiveness of buffer zones and corridors around and between reserves.

Monitoring remnant communities and the effects of insularization

Monitoring species and populations on islands is comparatively simple compared with monitoring species and populations in fragmented habitats. This is because the product of insularization is subjected to invasions, succession and human impacts. One difficulty of monitoring populations in remnant habitats (highlighted by Wilcox & Murphy 1985) is that insularization results in habitat fragments which may lack microhabitat diversity necessary for population persistence or stability. That 'atypical' phenologies may be found amongst those populations in remnant habitats lacking microhabitat diversity has been confirmed in a study of indicator nymphaline butterfly species found in patches of grassland in the San Francisco area (Ehrlich & Murphy 1987). They found for example that those *Eupydryas editha* populations which resided on solely east-facing slopes had delayed larval development times, emergence curves and peak-flight periods as well as shortened periods of larval host and adult nectar source availability. Such conditions affected daily sex ratios and resident times compared to those areas which supported a variety of habitats.

The importance and value of monitoring populations in remnant communities as well as the difficulties have been discussed in detail by Ehrlich & Murphy (1987). They have suggested that such monitoring programmes can contribute towards an understanding of population dynamics and processes. They also emphasize the value of studies on population genetics because of the important application of such research in the design of management strategies for isolated populations.

Small isolated populations may be subjected to genetic drift and inbreeding which diminishes genetic variability and then limits the extent to which the population can adapt to changing environmental conditions. Miller (1979) has suggested that in East Africa, changing environmental conditions could affect the ability of small populations of large mammals to adapt successfully to periodically recurring extreme dry or wet conditions. In instances where populations of endangered species have been reduced or are isolated in reserves, it would seem important to monitor not only populations levels and recruitment but also the genetic variability. A genetic basis for species vulnerability and a case for monitoring has been, for example, described for the Cheetah (*Acinonyx jubatus*) by O'Brien *et al.* (1985). High genetic uniformity has been found in South African Cheetahs and one population was found to contain 10 to 100 times less genetic variation than other mammal species. It is possible that a severe population bottleneck has occurred in this species in recent evolutionary times. Such genetically uniform species, especially in low population densities, have a bleak future unless new sources of genetic variation can be introduced into breeding programmes (O'Brien *et al.* 1985). Monitoring the status of the Cheetah would therefore require a population basis as well as a genetic basis.

The role of nature reserves as monitoring sites

The creation of nature reserves has been one response to the reduction and disturbance of natural communities. Nature reserves have not only been seen as last havens for wildlife but have also been considered to be important locations where natural processes, least affected by human disturbance, can be researched. The use of nature reserves and other protected natural areas for ecosystem monitoring and long-term ecological studies is not new (Jenkins & Bedford 1973, Cairns 1981) but few countries seem to have initiated monitoring programmes in protected areas. The USA is one country where there is monitoring and long-term data collection as part of the UNESCO Biosphere Reserves programme (White & Bratten 1981).

In addressing the great importance of conservation of biological diversity, the IUCN (IUCN 1989) has recently recommended in the report 'From Strategy to Action' that there be an establishment of a global system of Biological Diversity Monitoring Stations. Perhaps at least some of those stations could be located on protected areas.

References

Bernstein, I. S., Balcaen, P., Dresdale, L., Gouzoules, H., Kavanagh, M. & Patterson, T. (1976). Differential effects of forest degradation on primate populations. *Primates*, **17**, 401–11.

Brakefield, P. M. (1991). Genetics and the conservation of invertebrates. In *The Scientific Management of Temperate Communities for Conservation*, ed. I. F. Spellerberg, F. B. Goldsmith & M. G. Morris, pp. 45–79, Oxford, Blackwells.

Burley, F. W. (1988). Monitoring biological diversity for setting priorities in conservation. In *Biodiversity*, ed. E. O. Wilson & F. M. Peter, pp. 227–30, Washington, National Academy Press.

Butcher, G. S., Niering, W. A., Barry, W. J. & Goodwin, R. H. (1981). Equilibrium biogeography and the size of nature preserves. *Oecologia*, **49**, 29–37.

Cairns, J. (1981). Biological monitoring Part VI – future needs. *Water Research*, **15**, 941–52.

Darby, H. C. (1956). The clearing of the woodland in Europe. In *Man's Role in Changing the Face of the Earth*, ed. W. L. Thomas, pp. 183–216, Chicago, University of Chicago Press.

Ehrlich, P. R. & Murphy, D. D. (1987). Monitoring Populations on remnants of Native Vegetation. In *The Role of Remnants of Native Vegetation*, ed. D. A. Saunders, G. W. Arnold, A. A. Burbridge & A. J. M. Hopkins, pp. 201–10.

Fonseca, G. A. B. (1985). The Vanishing Brazilian Atlantic Forest. *Biological Conservation*, **34**, 17–34.

Green, G. M. & Sussman, R. W. (1990). Deforestation history of the eastern rain forests of Madagascar from satellite images. *Science*, **248**, 212–15.

Huntley, B. J. (1988). Conserving and monitoring biotic diversity. Some African examples. In *Biodiversity*, ed. E. O. Wilson & F. M. Peter, pp. 248–60, Washington, National Academy Press.

IUCN (1989). *From Strategy to Action. The IUCN response to the Report of the World Commission on Environment and Development*. Gland, IUCN.

Jenkins, R. E. & Bedford, W. B. (1973). The Use of Natural Areas to Establish Environmental Baselines. *Biological Conservation*, **5**, 168–74.

Lovejoy, T. E., Bierregaard, R. O., Rylands, A. B., Malcolm, J. R., Quintela, C. E., Harper, L. H., Brown, K. S., Powell, A. H., Powell, G. V. N., Schubart, H. O. R. & Hayes, M. B. (1986). Edge and other effects of isolation on Amazon forest fragments. In *Conservation Biology, The Science of Scarcity and Diversity*, ed. M. E. Soule, pp. 257–85. Massachusetts, Sinauer Ass.

252   Monitoring in practice

Lyster, S. (1985). *International Wildlife Law*. Cambridge, Grotius Publications Ltd.
Malingreau, J. & Tucker, C. J. (1988). Large-scale deforestation in the southeastern Amazon basin of Brazil. *Ambio*, **17**, 49–55.
Miller, R. I. (1979). Conserving the Genetic Integrity of Faunal Populations and Communities. *Environmental Conservation*, **6**, 297–304.
Newmark, W. D. (1987). A land-bridge island perspective on mammalian extinctions in western North American parks. *Nature*, **325**, 430–2.
O'Brien, Roelke, M. E., Marker, L., Newman, A., Winkler, C. A., Meltzer, D., Colly, L., Evermann, J. F., Bush, M. & Wildt, D. E. (1985). Genetic basis for species vulnerability in the Cheetah. *Science*, **227**, 1428–34.
Peterken, G. F. (1991). Ecological issues in the management of woodland nature reserves. In *The Scientific Management of Temperate Communities for Conservation*, ed. I. F. Spellerberg, F. B. Goldsmith & M. G. Morris, pp. 245–272. Oxford, Blackwells.
Repetto, R. & Gillis, M. (1988). *Public Policies and the Misuse of Forest Resources*. A World Resources Institute Book. CUP.
RSNC (1989). *Losing Ground, Habitat destruction in the UK: a review in 1989*. Nettleham, RSNC.
Spellerberg, I. F. (1981). *Ecological Evaluation for Conservation*. London, Edward Arnold.
Spellerberg, I. F. (1991). Biogeographical Basis of Conservation. In *Scientific Management of Temperate Communities for Conservation*, ed. I. F. Spellerberg, F. B. Goldsmith & M. G. Morris, pp. 293–322. British Ecological Symposium. Oxford, Blackwells.
Spellerberg, I. F. & Hardes, S. (1991). *Biological Conservation*. Cambridge, Cambridge University Press (in press).
Usher, M. B. (1988). Biological invasions of nature reserves: a search for generalizations. *Biological Conservation*, **44**, 119–35.
Verner, J., Morrison, M. L. & Ralph, C. J. (1986). *Wildlife 2000*. Madison, The University of Wisconsin Press.
Vujakovic, P. (1987). Monitoring extensive 'Buffer Zones' in Africa: an application for satellite imagery. *Biological Conservation*, **39**, 195–208.
WCED (1987). *Our Common Future*. Oxford, New York, Oxford University Press.
Webb, N. R. (1990). Changes on the heathlands of Dorset, England, between 1978 and 1987. *Biological Conservation*, **51**, 273–86.
Webb, N. R., Clarke, R. T. & Nicholas, J. T. (1984). Invertebrate diversity on fragmented *Calluna*-heathland: effects of surrounding vegetation. *Journal of Biogeography*, **11**, 41–6.
White, P. S. & Bratten, S. P. (1981). Monitoring vegetation and rare plant populations in US national parks and reserves. In *The Biological Aspects of Rare Plant Conservation*, ed. H. Synge, pp. 265–78. Chichester, John Wiley.
Whitney, G. G. & Somerlot, W. J. (1985). A Case Study of Woodland Continuity and Change in the American Midwest. *Biological Conservation*, **31**, 265–87.
Wilcove, D. S., McLellan, C. H. & Dobson, A. P. (1986). Habitat Fragmentation in the Temperate Zone. In *Conservation Biology, The Science of Scarcity and Diversity*, ed. M. E. Soule, pp. 237–56, Massachusetts, Sinauer Ass.
Wilcox, B. A. & Murphy, D. D. (1985). Conservation strategy: the effects of fragmentation on extinction. *American Naturalist*, **125**, 879–87.
Wildlife Link (1984). Loss and damage to SSSIs from agriculture and forestry operations, *Habitat Report*, **2**, London, Wildlife Link.

14
Monitoring land use and landscapes

Introduction

LAND is used for many purposes and conflicts between different land uses are becoming more common and more intense. The greatest diversity of land uses is possibly to be found in the industrialized countries where there is a desperate need for good information on land use and land cover as a basis for planning and resource assessment. The need for good information was recognized in Britain as early as the 1930s when Stamp (1962) launched the Land Use Survey which employed hundreds of volunteers to map every field and block of land. But such a basic land use inventory is rapidly outdated as a result of rapid changes in land use. However, mapping and classification of land has evolved over recent years to become a more highly sophisticated science. There are now many land use recording and monitoring programmes (both national and regional) for many types of features (ecological, land use, soil, land capability).

Whereas land use capability and physiography are 'static' classifications requiring only occasional surveillance, land use and ecological characteristics are continually changing at different rates and therefore need a more sophisticated basis for monitoring. That basis will depend on the aims of the monitoring whether it be for identification of water resource planning or the identification of ecologically sensitive areas.

There are, however, few examples of long-term land use monitoring programmes and this can be attributed to a lack of logistical and financial support for long-term programmes which leads to lack of uniformity in the variables being measured and lack of coordinated databases. Perhaps the CORINE Programme (see p. 48) will provide an incentive for the establishment of good land use monitoring programmes.

Anyone looking for data on land use may be surprised at the difficulties in obtaining published accurate data. This difficulty exists despite the contrary impression given by widely quoted figures on land uses and the publication of various land use statistics. In this section we consider very briefly the organizations involved, the problems encountered in monitoring land use and landscapes (especially in relation to ecological change and wildlife conservation), then describe some examples of land use monitoring programmes.

Organizations and land use data

Land use has been dominated by agriculture throughout the world and it is estimated that about 9 million sq km of the earth's surface has been converted into permanent croplands. It is relatively easy to calculate, in broad terms, a global figure such as this but it is far more difficult to determine accurately on a countrywide basis the extent and nature of change in a major land use such as agriculture. Some of these difficulties have been overcome by the US Geological Survey with LANDSAT data coupled with aerial photography and ground surveys. Since 1975, mapping and monitoring land use and land cover has been undertaken at a scale of 1:25 000 with the aim of producing baseline data every 6–7 years (Anderson 1977).

In Britain, data on land use come from widely varied sources and there is no one comprehensive source as Table 14.1 indicates. In addition to those sources mentioned in Table 14.1 there are Ordnance Survey maps and records held by local authority planning departments but those records may not contain the type of land use data needed in a particular situation. Reliable and accurate land use data are also dependent on research and development of appropriate methods to meet particular needs and scale of work. Work of this kind has been undertaken in a number of centres including the Land Use Research Unit at King's College, London (Coleman & Shaw 1980) and the Geodata Institute at Southampton University. The Sussex Land Use Inventory, a database recording land use in Sussex is also noteworthy in this respect (Lukehurst & Rimmington 1981).

Agriculture is the dominant land use in Britain and perhaps not surprisingly comprehensive records of agricultural land use are provided by the detailed agricultural statistics produced by the Ministry of Agriculture, Fisheries and Food (MAFF) annual census (the June returns). The 1983 census reported that 78 per cent of the total area of the UK was in agricultural use. The MAFF data are concerned primarily with land being used for agricultural production and do not mention agricultural land which has been designated for other reasons such as protected areas including Sites of Special Scientific Interest and Areas of Outstanding

Table 14.1. *Examples of land use and landscape surveillance information of relevance to ecology and conservation in Britain*

1. Department of the Environment.
   (i) Derelict Land Returns, 1974, 1982 census of derelict land.
   (ii) Digest of Environmental Protection and Water Statistics. Annual reports which include summary data from various sources on land use and nature conservation. London, HMSO.
   (iii) *National Land Use Classification*. HMSO (1975).
   (iv) Land use change in England. DOE Statistics Bulletins.
   (v) *National Land Use Stock Survey. A Feasibility Study for the Department of the Environment*, Roger Tym & Partners (1985). A two-volume report with excellent overviews on land use classification and monitoring and survey techniques. This report also contains data on land use from various agencies.

2. Department of the Environment and Countryside Commission.
   Monitoring Landscape Change. Hunting Technical Services Ltd. (1986). Full report in ten volumes and summary report by the Countryside Commission.

3. Department of Transport.
   Road lengths in GB 1986–87. Statistical Bull (88) 35 HMSO (1988).

4. Ministry of Agriculture, Food and Fisheries.
   Annual Agricultural Census. Concerned mainly with agricultural production and type of farming.

5. MAFF: Agricultural Development Advisory Service (ADAS).
   Agricultural Land Classification. A report based on a 1966–74 survey.

6. Forestry Commission.
   Woodland Census. Reports every 15 years and last report was 1980.

7. Institute for Terrestrial Ecology.
   Landscape Changes in Britain. ITE (1986) A survey of the rural environment based on field surveys in 1977–78 and 1986.

8. Nature Conservancy Council.
   National Countryside Monitoring Scheme.

9. Countryside Commission.
   New Agricultural Landscapes, CCP76. Cheltenham: The Commission. This and later similar reports have information on changes in agricultural land, extent of hedgerows, changes in field size, etc.

10. BBC.
    BBC Doomsday Project. A survey undertaken by 11 500 schools and based on subjective impressions of main land uses and land cover.

11. Coast watch schemes: Coastwatch (NCC) and Norwich Union Coastwatch UK.

Natural Beauty. The Forestry Commission's data are of a fairly specialized nature and are limited in application in monitoring because the data are collected approximately every 15 years. The Department of the Environment has published data on derelict land (from 1974 and 1982 surveys) and more recently, together with the Countryside Commission, has established a Landscape Monitoring Programme. The Nature Conservancy Council has regularly published information (NCC Annual Reports) about the extent and numbers of protected areas in Britain but more importantly has established a National Countryside Monitoring Scheme (NCC 1987, 1989; NCC/CCS 1988), the details of which are discussed on p. 269.

All of these inventories and monitoring programmes have been based on different land use classifications and different data storage and retrieval systems, some mapped base, some simply recording total acreage. Therefore it has been difficult to develop comprehensive, national land use statistics or land use maps, let alone comprehensive monitoring programmes. A National Land Use Classification Scheme was suggested by the DOE in 1974 for its initial attempt to collect land use change statistics from local authorities and this was later published in 1975 (Dickinson & Shaw 1978). However, problems in collation of data arose from the fact that different local authorities held their respective data under different category headings. In 1984, the DOE remarked that it 'intended to monitor land use change using data supplied by the Ordnance Survey, starting in January 1985.' This intention seems to have been prompted by a policy to protect the countryside for better food production but later this policy was changed to one of protecting the countryside for its own sake rather than primarily for the productive value of the land. A feasibility study of methods and costs was commissioned by the DOE and eventually a feasibility study (National Land Use Stock Survey) was published in 1985 (Tym & Partners 1985).

Collecting, storing and analysing data

Objectives of data collection

Land use data collection, analysis (and classification) is a vast subject (Dawson & Doornkamp 1973, Coleman 1985) and here we can only summarize the more important aspects as a background to descriptions of some land use and landscape monitoring programmes. The purpose of the land classification, whether it be for monitoring change or as a basis for assessment of an area's potential, will determine the methods used for data collection. The frequency of data collection will also dictate certain methods, that is some land attributes are continuously changing and therefore need repeated monitoring whereas some attributes such as soil type are static and would need only occasional data collection.

It is also useful to distinguish between land surveys used to map spatial patterns and other surveys which are designed to measure extent of land use in each class or category. We also need to distinguish between land mapping which is designed to tell us what is there now and those schemes designed to assess potential value or potential use of the land (capability maps and surveys; see examples in Dawson & Doornkamp).

Sources of data

Sources of data for monitoring land use and land cover can be either secondary (e.g. Ordnance Survey maps) or primary and collected specifically for the purposes of monitoring. Primary data for monitoring land use and land type change has previously been obtained from ground surveys, aerial photography and satellite imagery (see, for example, p. 175). Ground surveys or inspection of the land on foot potentially allow as much land to be examined in as much detail as time and facilities permit. Such systematic recording has considerable advantages in terms of accuracy but may be costly in terms of time and effort. Aerial photography has long been used to study changes in landscapes and land uses and this method has been well researched (Barrett & Curtis 1974, Clarke 1986). In practice, aerial photography requires careful interpretation and by itself can't provide detailed information about ecological attributes such as the interior structure of a woodland or the species composition of grassland. Nevertheless such techniques are suitable for less structurally complex vegetation such as coastal dune vegetation. Remote sensing techniques with satellite imagery have greatly improved the accuracy of recording and monitoring land use change and have widened the application of landscape and land use monitoring but nevertheless the success of satellite imagery is dependent on acquisition of good ground reference data.

In most cases, the extent of land being classified and monitored would be so great that comprehensive coverage at regular intervals for monitoring would be impractical. In the absence of comprehensive coverage there has to be a method of data collection by sampling units of which there are three broad methods: area, line and point sampling. Line sampling simply uses points along a line between two locations as a basis for recording land uses and land cover, a technique used in surveys of area but not spatial patterns. Point and area sampling are similar in practice but points have a location and no area. In both methods, land use and land cover are recorded at a point or in a sample area.

The importance of giving some care to the choice of number of sampling units and the resolution of spatial sample units can be illustrated with reference to ecological features such as woodlands. There are many shapes and sizes of woodlands scattered throughout the countryside, some as small as 0.1 ha. Depending on the frequency and distance between points used for sampling, some woodlands may fall between points or the proportion of land sampled could also fail to record certain types of woodlands in some localities. Similarly, the spatial distribution of rare plant species or rare animals in a community may not be detected by the sampling methods adopted. Land cover surveys may therefore require additional and more detailed surveys in some parts of the region being considered.

Since 1972, the LANDSAT series of satellites with systematic and regular coverage of the earth's surface have been a major source of data for use in analysis of land use mapping and monitoring programmes. The two principal sensors, the Multispectral Scanner and the more recent Thematic Mapper have been used widely for monitoring changes in medium- to large-scale features and there is now much research being carried out to try and extend the use of satellite imagery for use in monitoring small features. The NERC Unit for Thematic Information Systems (NUTIS) has, for example, been assessing SPOT (Systeme Probatoire d'Observation de la Terre – based on satellites placed in orbit in the mid-1980s) for its use in monitoring land use change. In the USA, both NASA and the National Academy of Sciences have shown an interest in remote sensing using satellites on a global scale, particularly in connection with the proposed International Geosphere Biosphere Programme (Tuyahov, Star & Estes 1989).

There have been many applications of satellite technology for monitoring environmental changes, particularly broad-scale changes. Vujakovic's (1987) work using satellite imagery for studying the use of buffer zones in relation to conserving the ecological integrity of protected areas in Botswana has already been mentioned (Chapter 13). In that example, classification of semi-arid vegetation based on LANDSAT MSS Band 5 was achieved with relatively high accuracy.

Data storage and analysis

Mapping of ecological variables, using a range of techniques, is popularly called ecological mapping but that is a simplification. Ecological mapping is indeed concerned with mapping of organisms, assemblages and habitats (whether it be from the ground, air or space) but it can also be concerned with mapping fungi on a forest floor or microbes in the soil (Fuller 1983). Ecological mapping has become a sophisticated science in the Netherlands for the identification and monitoring of areas of conservation and ecological importance (see, for example, Kalkhoven, Stumpel & Stumpel-Rienks 1976). Ecological mapping of coastal zones to enable the identification of ecologically sensitive areas has also been developed in connection with assisting in responses to pollution incidents such as oil spills. With this kind of information it is theoretically possible to direct clean-up operations so that they are most effective in those areas in greatest need of protection. These same data also form part of baseline information for ecological monitoring.

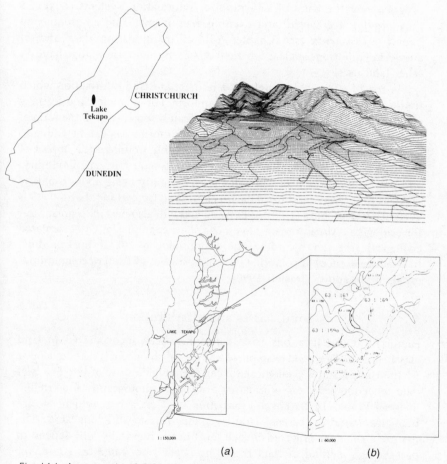

(a)   (b)

Fig. 14.1. An example of GIS potential: this shows a digital terrain model for Lake Tekapo in the South Island of New Zealand. Ecological units are shown viewed from 3000 m altitude due east from the lake. From Espie & Hall (1989) with kind permission of the authors.

Spatial distribution data have traditionally been presented in map form but this becomes cumbersome to handle especially when several categories of information (such as soil characteristics, water characteristics, current land uses, land capability) are being analysed. Recent advances in the sophistication of computers and computer programs have greatly influenced the way we handle, store, analyse and present data. Computerized databases are becoming a more and more important and integral part of biological monitoring and so too are computers that store spatial and descriptive data about objects and the relationship of those objects to other objects on the earth's surface. Those computers and the computer programs are the basis of Geographical Information Systems (GIS) and they provide a powerful and exciting way of analysing environmental impact assessments (see Chapter 15), of manipulating data and of answering questions such as what effect will this land use change have on other land uses?

The use of GIS enables interaction of databases and information which traditionally had been presented in map form. For example, land features such as biological communities or habitats can be portrayed in the form of digital terrain models (Fig. 14.1). Additional information can be added to the model as required; for example, it's possible to follow the impact of planting a conifer plantation or constructing a reservoir. The flexibility, and powerful capacity, of GIS will almost certainly bring about dramatic changes in the way land use data are monitored (something that NUTIS is currently researching (NERC 1989)) and GIS will also become a great asset for ecologists. Already there have been some examples of GIS-orientated ecological research including the phytosociology of alpine meadow, habitat preferences of passerine birds and studies of Diptera communities on alpine meadows (Haslett 1990).

Land classes and classification

Hierarchical classifications have commonly been incorporated in land classifications. One good example of a hierarchical system is that developed by the United States Geological Survey which makes use of two levels of data collection based on information from aerial photography and satellite imagery (Table 14.2). Level 1 has nine land classes of which six are biological: tundra, 'barren' land, wetland, forests and rangeland. The nature of the data collected at each level in the hierarchy will depend in part on the method of data recording. Data from satellites could, for example, identify areas of tundra (high-latitude or high-altitude treeless regions dominated by low shrubs, lichens, mosses and sedges), aerial photography could identify categories at the next level such as bare ground tundra, wet tundra or herbaceous tundra. Ground surveys could

Table 14.2. *United States Geological Survey land use and land cover classification for use with remote sensing*

Level 1	Level 11
1. Urban or built land	11 Residential
	12 Commercial and services
	13 Industrial
	14 Transportation, communication and utilities
	15 Industrial and commercial complexes
	16 Mixed urban or built-up land
	17 Other urban or built-up land
2. Agricultural land	21 Cropland and pasture
	22 Orchards, groves, vinyards, nurseries and ornamental horticultural areas
	23 Confined feeding operations
	24 Other agricultural land
3. Range land	31 Herbaceous rangeland
	32 Shrub-brushland, rangeland
	33 Mixed rangeland
4. Forest land	41 Deciduous forest land
	42 Evergreen forest land
	43 Mixed forest land
5. Water	51 Streams and canals
	52 Lakes
	53 Reservoirs
	54 Bays and estuaries
6. Wetland	61 Forested wetland
	62 Nonforested wetland
7. Barren land	71 Dry salt flats
	72 Beaches
	73 Sandy areas other than beaches
	74 Bare exposed rock
	75 Strip mines, quarries and gravel pits
	76 Transitional areas
	77 Mixed barren land
8. Tundra	81 Shrub and brush tundra
	82 Herbaceous tundra
	83 Bare ground tundra
	84 Wet tundra
	85 Mixed tundra
9. Perennial snow or ice	91 Perennial snowfields
	92 Glaciers

From US Geological Survey Professional Paper No. 964 (Appendix A).

extend the hierarchy to another level to identify species associations within wet tundra or herbaceous tundra.

Land classification schemes can be used as a basis for evaluating the potential of land for one or more uses or they can be used to tell us what is present. Evaluation is a form of classification and may be undertaken in relation to identification of the potential of land for many purposes such as mineral resources, agriculture or forestry, production of wood for fuel, or for settlement (Young 1973). Classification schemes used to identify what is there might be directed at soil types, physiographic features, aquatic habitats or woodland types. Land classification has to be based on a suitable choice of land units or regions and the choice of those land units or regions will depend on the objectives of the land classification. For example, a hypothetical woodland could be classified as a mixed deciduous woodland or as a recreational area or both, depending on the woodland's attributes and use.

A draft standard international classification of land use was published in 1983 by the Statistical Commission and Economic Commission for Europe. Developed in close cooperation with international organizations such as FAO and OECD, this classification has been established for the purposes of physical planning and land management. Both national uses and international uses are envisaged because of concerns about the need for information on current land use, demands for land resources, environmental repercussions of land use and planning of future developments of land use patterns. This draft standard has seven first-level categories based on physical characteristics, i.e. agricultural land, forest and other wooded land, built-up and related land, open wetlands, dry open land with special vegetation cover, open land without, or with insignificant, vegetation cover, and water. At other levels, functional characteristics are used for the classification. No size limit is applied for classification of land apart from a minimum area of half a hectare for agricultural land.

Data from Ordnance Survey maps and other maps have been used successfully by the Institute for Terrestrial Ecology (ITE) in Britain for land classification and resource assessment based on the 1 km² National Grid (Smith 1982, Bunce & Heal 1984). Data collection and data analysis included two stages: 1, the grouping of grid squares into arbitrary land classes; and 2, the characterization of each land class by field surveys.

The grouping of grid squares into land classes was based on the multivariate classification (Indicator Species Analysis (ISA), Hill 1979) of data available from Ordnance Survey and geological maps. The complex of geological, topographical, and functional map attributes within each grid square characterize each land class. Indicator species analysis is a divisive polythetic method of numerical classification which has been bound to be applicable to large sets of either qualitative or quantitative data. Details of the method may be found in Hill, Bunce & Shaw (1975, p. 612 *et seq.*). More recently the advent of TWINSPAN has facilitated such techniques and their general use for attribute data.

Although samples of grid squares were used (usually the centre grid in each set of nine) Indicator Species Analysis produces an efficient key to the classification, enabling allocation of all grid squares to the most appropriate land class. Obviously the interpretation of land classes depends upon the main criteria selected by the indicator analysis. An example of the land-class map characteristics, appropriate for Cumbria in the northwest of England (where the main project was originally based) is shown in Table 14.3 and it can be seen that these classes are not analogous to ecological categories such as woodland or moorland. This is because each land class is used as a strategy for subsequent random selection of 1 km² areas for surveying a region. Each class therefore contains a range of ecosystems.

In the second stage, data collected in the field from a random selection of grid squares enable each land class to be characterized. These data can then be used to identify ecological characteristics or land use. The distribution of those characteristics is based on the land-class distribution. The level of definition such as distribution depends on the grid size. Choice of grid size ultimately depends upon the level of detail required in the predicted distribution patterns and upon resources available for the classification and field surveys. Grid squares ranging from 0.25 km² to 10 km² have been used by the ITE, and Bunce has used 10 km² for an ecological survey of Britain (Bunce *et al.* 1984).

A test of the accuracy of this method can be achieved by comparing observed characteristics (from a sub-sample of grid squares) with the expected characteristics within a sub-region. On a larger scale, land classes can be used for estimates of countrywide land use and the results can then be compared for accuracy with other sources of data. For example, Bunce & Heal (1984) have reported that estimates of land use based on this system of land use classification correspond closely with data from other sources. The level of accuracy depends on the size of the area for which information is sought and the scale to which the sample field surveys are based. Estimates of the ecological characteristics of a region based on the national sample surveys of the 32 ITE land classes defined nationally become more accurate as the size of the region decreases. The land classification schemes, such as that described above, can also provide a basis for predictive modelling. The models being developed for predicting changes in rural land use in Britain now aim to incorporate socio-economic, ecological and hydrological information, all based on the common 1 km² grid land classification developed by Bunce and others.

Table 14.3. Map characteristics for 16 Cumbrian land classes

Land class	Constant (80–100%) characters	9 most important characteristic features				Characteristics which help distinguish classes – 1 from 2, 3 from 4, 5 from 6, etc.	
7	0–249 ft Sea	Intertidal Sand and mud	Sand and mud Sea	Intertidal Marsh	0–249 ft Basin peat	Basin peat River gravels	Alluvium Sand and mud
8	Intertidal 0–249 ft	Sea	Intertidal Sea Sand and shingle	Sand and mud 0–249 ft Railway	Sandstone	Grey house Railway	White road
6	0–249 ft Grey house	Yellow road Basin peat White road	Embankment Town 0–249 ft	Railway (disused) Basin peat Sandstone	'A' road Church Bridleway	'A' road Cutting Railway (disused)	Embankment River
5	0–249 ft White road		0–249 ft Sandstone	Basin peat White road	Basin peat Grey house	0–249 ft	
1	Grey house White road	Yellow road Footpath	250–499 ft 0–249 ft Grey house	White road Yellow road 'A' road	Copse Footpath Black house	0–249 ft	
2	Grey house 250–499 ft	Yellow road White road	250–499 ft Penrith sandstone Yellow road	Grey house River White road	Black house Sandstone Footpath	Kirklington/St Bees sandstone Penrith sandstone	River 500–749 ft
4	Stream 500–749 ft	Yellow road Grey house	500–749 ft Bannisdale slates Yellow road	Unfenced road Hamlet Grey house	River Bracken/heath Copse	Bannisdale slates Bracken/heath	500–749 ft
3	White road	Black road	Limestone	Black road	Unfenced	Limestone	

Site						
11	Bracken/heath Stream	750–999 ft Bracken/heath	500–749 ft Bannisdale slates	Bracken/heath Limestone	Bannisdale slates Copse	Aspect (N)
12	Bracken/heath	Fell sandstone Slope	750–999 ft Wood (conifer)	Unfenced road 1500–1749 ft	Fell sandstone	Slope
10	Stream 1000–1249 ft	1000–1249 ft 1250–1499 ft	750–999 ft Bracken/heath	1500–1749 ft Borrowdale volcanics	Black house Grey house Limestone	Unfenced road White road
9	1000–1249 ft Bracken/heath	1000–1249 ft 1250–1499 ft	Bracken/heath 750–999 ft	Bracken/heath Limestone		
13	1500–1749 ft Stream	1500–1749 ft 1750–1999 ft	Millstone grit 1250–1499 ft	Limestone	1250–1499 ft	1500–1749 ft
14	Stream Bracken/heath	2000–2249 ft 2250–2749 ft	1750–1999 ft Bracken/heath	1000–1249 ft	Spot height 2000–2249 ft	2250–2499 ft
15	1500–1749 ft Stream 1250–1499 ft	1500–1749 ft 1250–1499 ft 1750–1999 ft	Borrowdale volcanics National Trust property Bracken/heath	National Trust property 1500–1749 ft	1250–1499 ft	
16	2000–2249 ft Bracken/heath Borrowdale volcanics Stream Aspect (W) 1750–1999 ft Scree/crag Aspect (N)	2000–2249 ft 2250–2499 ft 2500–2749 ft Scree/crag	1750–1999 ft	Bracken/heath	Scree/crag 2000–2249 ft	2250–2499 ft 2500–2749 ft

From Smith (1982).

Land use and land cover monitoring programmes

The following are examples of land use and landscape monitoring programmes. Some are rather crude and others are sophisticated with potentially wide and useful applications. There is, however, much concern expressed at the lack of uniformity in data classes, types of data acquired and the computing methods employed by those programmes in Britain. The history of the various land use monitoring programmes in Britain would also seem to suggest that there has been minimal communication between organizations.

DOE/CC Project

There has been much concern expressed about the lack of reliable and accurate information about landscape changes in England and Wales. For this reason a research project was commissioned by the Department of the Environment and the Countryside Commission. Using sources of data from aerial photography and satellite imagery backed up by ground surveys, the aim of the research was to 'assemble statistical information on a consistent basis on the current and past distribution of landscape features of major policy importance and on the changes in distribution that had taken place in the post-war period' (DOE 1988). The full report (Department of the Environment and Countryside Commission 1986), *Monitoring Landscape Change*, in ten volumes was produced for the DOE and the Countryside Commission by Hunting Technical Services Ltd. but summaries of this lengthy report may be found in DOE publications (DOE 1988, 1989) and Countryside Commission publications.

The DOE/CC Project included two separate but linked studies: one looking at the extent and distribution of linear features such as hedges and walls in 1947, 1969, 1980 and 1985, the other study looking at the extent of land form and land use categories such as woodland, semi-natural vegetation, agricultural land and water for the years 1947, 1969 and 1980. Estimates of extents of various categories were obtained from aerial photography (supplemented by some ground surveys) and from the LANDSAT Thematic Mapper.

Two examples of the results from the DOE/CC project are shown in Tables 14.4 and 14.5. In addition to the DOE and the Countryside Commission, other organizations have recorded and published changes in some of these landscape categories and it is interesting to find not unexpected differences in the figures. For example, the DOE/CC project reports a loss of about 4000 to 5000 km of hedgerows per year during the period 1947–1985. The ITE estimates of hedgerow loss for the period 1978–1984 imply slightly lower net losses (ITE 1986) but a much lower estimate was made by MAFF in 1985.

Table 14.4. *Landscape changes in England and Wales: changes in the extent of features from 1947 to 1980*

Landscape features	1947[a] % cover	1947[a] Standard error	1969[a] % cover	1969[a] Standard error	1980[a] % cover	1980[a] Standard error
Broadleaved woodland	5.6	0.3	4.7	0.3	4.2	0.2
Coniferous woodland	0.7	0.2	2.2	0.4	2.8	0.5
Mixed woodland	0.7	0.1	1.0	0.1	0.9	0.1
All woodland	7.0	0.4	7.9	0.5	7.9	0.6
Upland heath	3.0	0.6	2.4	0.6	2.4	0.6
Upland grass (smooth)	1.2	0.3	0.9	0.2	0.6	0.2
Upland grass (coarse)	4.6	0.7	4.3	0.7	3.9	0.7
Bracken	1.1	0.2	1.0	0.2	1.0	0.3
Lowland grass heath	1.5	0.2	0.4	0.1	0.3	0.1
All semi-natural vegetation[b]	12.6	0.9	10.1	1.0	9.2	0.9
Cultivated land	26.1	0.8	31.7	0.8	35.4	0.9
Improved grassland	38.1	1.1	34.5	1.1	31.0	0.9
Rough grassland	2.9	0.3	2.2	0.2	2.2	0.2
Neglected grassland	3.7	0.3	3.6	0.3	3.1	0.3
All farmed land	72.7	1.1	72.1	1.0	71.8	1.0
All water and wetlands[c]	1.3	0.1	1.1	0.2	1.1	0.2
Built-up land[d]	4.5	0.3	6.5	0.3	7.3	0.4
Urban open space[d]	0.7	0.1	1.1	0.1	1.3	0.1
All other land[e]	6.4	0.4	8.8	0.4	9.9	0.5
All features	100.0		100.0		100.0	
Area (000 km)	151.1		151.1		151.1	

[a]Photographs were interpreted for a range of dates around this year.
[b]Includes blanket bog, lowland heather and gorse.
[c]Coastal and inland waters, freshwater marsh, salt marsh and peat bog.
[d]Urban land can be defined in many ways. For example, Professor Best estimated that urban land defined as covering all land under urban uses covered the following proportion of England and Wales: 6.7 per cent in 1930, 8.8 per cent in 1950, 10.8 per cent in 1970 and 11.6 per cent in 1980.
[e]Includes major transport routes, bare rock, sand, shingle, mineral works and derelict land.

From DOE (1989). Reproduced with the permission of the Controller of Her Majesty's Stationery Office.

Table 14.5. *Landscape changes in England and Wales; changes in linear features*

	Landscape feature											
	Hedgerows		Fences		Banks		Open ditches		Walls		Woodland fringe	
	Length	Standard error	Length	Standard error	Length	Standard error	Length	Standard error	Length	Standard error	Length	Standard error
1947[a]	796	53	185	10	151	27	122	35	117	30	241	20
1969[a]	703	51	193	11	140	26	116	33	114	30	241	20
1980[a]	653	49	199	11	132	24	111	32	111	29	243	20
1985	621	48	210	11	128	23	112	32	108	28	243	20

[a]Photographs were interpreted for a range of dates around this year.
From DOE (1989). Reproduced with the permission of the Controller of Her Majesty's Stationery Office.

NCC: National Countryside Monitoring Scheme

The Nature Conservancy Council has established a National Countryside Monitoring Scheme (NCMS) to record and examine changes in the distribution and extent of 'structural components' of the countryside since the 1940s, especially important conservation and ecological attributes such as woodlands, moorland and wetlands (the classification includes only one category of built land). The objective is to establish a standard and technically robust scheme for providing quantitative data about habitats and the changes they have undergone.

The NCMS was directed first at Cumbria (northwest England) then later in the Grampians of Scotland. For Cumbria, the first step was a stratification based on one of the 16 land classes (see Table 14.3). For this project these land classes were grouped into three broad land types (lowland, intermediate and upland). To ensure adequate sampling of the three types, each 5 km × 5 km square was allocated to a land type on the basis of the land type to which the most 1 km × 1 km squares within it had been assigned (NCC 1987, 1989).

Aerial photographs taken between 1945–49 and then later between 1970–76 provided the source of the data for estimating areas and changes in areas. The estimates were calculated by multiplying up the results from sampled areas, using the area covered by each kind of feature within each land class. The results were then presented as total estimates for Cumbria and for the three broad land types.

Grassland was found to be dominant (about 62 per cent) in both the 1940s and the 1970s followed by moorland habitats covering about 18 per cent of Cumbria in the 1940s and decreasing to about 12 per cent in the 1970s. Coniferous plantations were found to have increased by as much as 127 per cent. The percentage net loss for hedgerows was found to be about 38 per cent for the uplands and 22 per cent for the lowlands.

These data show that quite dramatic changes have taken place between the two periods and the major causes of those changes can easily be identified. Although this is valuable information, such monitoring done in retrospect and based on two sets of aerial photographs has its limitations, one being that it can not tell us anything about the rate and nature of insularization, a phenomenon which is more serious for wildlife conservation than simple loss of habitats (see Chapter 13). Accurate and reliable interpretation of black and white aerial photographs also has its limitations, not the least of which is the variation in the features depicted as shades of grey.

It is to be hoped that in the future a combination of foresight (identifying the need to establish long-term ecological monitoring), recent developments in remote sensing (using data gathered via satellites) and GIS (for sophisticated data handling) will provide a basis for monitoring land use and landscapes.

References

Anderson, J. R. (1977). Land use and land cover changes – a framework for monitoring. *Journal of Research US Geological Survey*, 5, 143–53.

Barrett, E. C. & Curtis, L. F. (1974). *Environmental Remote Sensing: Applications and Achievements*. London, Edward Arnold.

Bunce, R. G. H. & Heal, O. W. (1984). Landscape evaluation and the impact of changing land-use on the rural environment: the problem and approach. In *Planning and Ecology*, ed. R. D. Roberts & T. M. Roberts, pp. 164–88. London, Chapman & Hall.

Bunce, R. G. H., Tranter, R. B., Thompson, A. M. M., Mitchell, C. P. & Barr, C. J. (1984). Models for predicting changes in rural land use in Great Britain. In *Agriculture and the Environment*, ed. D. Jenkins, pp. 37–44, (ITE Symposium No. 13). Abbots Ripton, Institute of Terrestrial Ecology.

Clarke, R. (ed.) (1986). *The Handbook of Ecological Monitoring*. Oxford, Clarendon Press.

Coleman, A. (1985). Lessons from comprehensive national ground survey. State of the Art review: Methods of data capture 1. In *National Land Use Stock Survey. A Feasibility Study for the Department of the Environment*, ed. R. Tym & Partners, Appendix 9, pp. 121–44. London, Roger Tym and Partners.

Coleman, A. & Shaw, J. E. (1980). *Land Utilisation: Field Mapping Manual Second Land Utilisation Survey*. King's College London.

Dawson, J. A. & Doornkamp, J. C. (1973). *Evaluating the Human Environment*. London, Edward Arnold.

Department of the Environment and Countryside Commission (1986). *Monitoring Landscape Change*, 10 vols. Huntingdon Technical services, Borehamwood.

Dickinson, G. C. & Shaw, M. G. (1978). The collection of national land use statistics in Great Britain: a critique. *Environment and Planning*, 10, 295–303.

DOE (1988). *Digest of Environmental Protection and Water Statistics*, No. 10. London, HMSO.

DOE (1989). *Digest of Environmental Protection and Water Statistics*, No. 11. London, HMSO.

Espie, P. & Hall, G. (1989). Using a geographical information system for analysis of ecological data. In *Management of New Zealand's Natural Estate*, ed. D. A. Norton, pp. 31–5. New Zealand Ecological Society Occasional Publication No. 1.

Fuller, R. M. (1983). *Ecological Mapping from Ground, Air and Space*. Cambridge, Institute of Terrestrial Ecology.

Haslett, J. R. (1990). Geographic information systems: a new approach to habitat definition and the study of distributions. *TREE*, 5, 214–18.

Hill, M. O. (1979). *TWINSPAN – a FORTRAN Program for Arranging Multivariate Data in an Ordered Two-way Table by Classification of the Individuals and Attributes*. Section of Ecology and Systematics, N.Y., Cornell University.

Hill, M. O, Bunce, R. G. H. & Shaw, M. W. (1975). Indicator species analysis: a devisive polythetic method of classification and its application to a survey of native pinewoods in Scotland. *Journal of Ecology*, **63**, 597–613.

ITE (1986). *Landscape Changes in Britain*, Huntingdon, ITE.

Kalkhoven, J. T. R., Stumpel, A. H. P., Stumpel-Rienks, S. E. (1976). *Landelijke Milieukartering, Environmental Survey of the Netherlands*. A landscape ecological survey of the natural environment in the Netherlands for physical planning on a national level. Rijksinstituut voor Natuurbeheer (Research Institute for Nature Management) No. 9, Staatsuitgeverij -'s-Gravenhage.

Lukehurst, C. T. & Rimmington, H. P. B. (1981). *The Sussex Land Use Inventory*. Coding manual. Brighton Polytechnic Press.

NCC (1987, 1989). Changes in the Cumbrian countryside. *Research and Survey in Nature Conservation* No. 6, Peterborough, NCC.

NCC/CCS (1988). *National Countryside Monitoring Scheme, Scotland, Grampian*. Countryside Commission for Scotland and the Nature Conservancy Council.

NERC (1989). NERC *Unit for Thematic Information Systems, Report for 1987/88*. NERC Unit for Thematic Information Systems, Department of Geography, University of Reading.

Smith, R. S. (1982). *The Use of Land Classification in Resource Assessment and Rural Planning*. Cambridge, Institute of Terrestrial Ecology.

Stamp, L. D. (1962). *The Land of Britain: its Use and Misuse*. London.

Tuyahov, A. J., Star, J. L. & Estes, J. E. (1989). Observing the earth in the next decades. In *Changing the Global Environment, Perspectives on Human Involvement*, ed. D. B. Botkin, M. F. Caswell, J. E. Estes & A. A. Orio, pp. 285–304. London, New York, Academic Press.

Tym, R. (1985). *National Land Use Stock Survey. A Feasibility Study for the Department of the Environment*, Two Vols. London, Roger Tym and Partners.

Vujakovic, P. (1987). Monitoring extensive 'buffer zones' in Africa: an application for satellite imagery. *Biological Conservation*, **39**, 195–207.

Young, A. (1973). Rural land evaluation. In *Evaluating The Human Environment*, ed. J. A. Dawson & J. C. Doornkamp, pp. 1–33. London, Edward Arnold.

15
Environmental Impact Assessments and monitoring

Introduction

AN ENVIRONMENTAL IMPACT ASSESSMENT is a formalized procedure used to assess the impacts (whether good or bad) of a development or project on various aspects of human welfare and the environment. An Environmental Impact Assessment (EIA) provides information which can be used to determine whether or not various aspects of a development or project conform with statutory requirements or are simply perceived as being environmentally acceptable. If not acceptable then alternatives or modifications to various aspects of the development may have to be considered. Recent texts with detailed accounts of EIAs include Clark & Hetherington (1988) and Wathern (1988).

Although the EIA process was developed in the 1960s many countries now have legislation requiring an Environment Impact Assessment prior to the commencement of any major development. The rationale underlying EIAs is to ensure that the development will not infringe on human health and safety and that the development will not cause damage to the natural and physical environment. An EIA also provides a basis for looking at alternative sites, assessing the effects of different scales of developments and minimizing actions which may affect human health and safety. Monitoring effluents at source is usually a requirement of an industrial development but the effects of toxic effluents on plants and animals (especially non-domestic or non-economic organisms) are rarely monitored. In addition, biological monitoring is not usually mentioned explicitly in EIA legislation or guidelines.

One interesting exception can be found in the definitive proposals for EIAs in Antarctica. As has been noted in Chapter 1, limited environmental protection has been afforded the Antarctic by the Antarctic Treaty.

272

However, proposals from the Scientific Committee on Antarctic Research (SCAR) finally led in 1987 to the adoption by an Antarctic Treaty consultative meeting of guidelines for a formal EIA procedure to be used in the Antarctic. Those guidelines include 'identification of measures, including monitoring programmes, that could be taken to minimize or mitigate impacts and detect unforeseen effects'.

Despite apparent widespread agreement amongst environmental biologists that biological and ecological monitoring should be part of the post-EIA process, there are few examples of post-EIA monitoring (Rosenberg *et al.* 1981, Westman 1985, Wolfe 1988). This lack of monitoring may have come about partly as a result of a lack of demands for monitoring in legislation. It is interesting to note, however, that even in the absence of legal requirements, some industrial organizations (for example BP, Shell, Philips and Britain's Central Electricity Generating Board; the last of these is now split into three companies) have established post-EIA biological monitoring programmes, some of which have proved to be of value in relation to environmental restoration and habitat renewal (Cowell 1978, Cowell & Monk 1981).

The aim here is to provide an outline of the EIA process and then consider some examples of developments which have, for various reasons, included biological monitoring in that process. The introduction of EIAs has greatly influenced our thinking about the role of biological monitoring and for that reason it seems useful to know something about the history of EIAs.

History and nature of EIAs

Although the term EIA is now more than 20 years old, basic screening processes for environmental impacts arising from developments have been undertaken over a much longer time. In the United States, the National Environment Policy Act (NEPA) passed by the United States Congress in 1969 required federal agencies to

include in every recommendation or report on proposals for legislation and other major Federal actions significantly affecting the quality of the human environment a detailed statement ... on the effects of that action.

It was that Act which first brought in a requirement for EIAs for federally funded projects in the USA. Over 17 000 EIAs have now been prepared in the USA and there has been much debate about the requirements of EIAs (Haug *et al.* 1984) and the effectiveness of EIAs (Roberts & Roberts 1984).

Since the early 1970s many countries, including Australia, Canada,

France, Ireland and the EEC have introduced EIA legislation despite much criticism to the effect that existing legislation was adequate for EIAs. Westman (1985) gives a comprehensive list of the status of EIAs in selected countries.

In Britain, the foundations for environmental assessment came with a planning system dating back to the 1947 Town and Country Planning Act. The DOE initially resisted any introduction of EIA legislation but in 1981 an Environment Sub-committee of the European Community recommended that EIAs had a role to play in Britain and accordingly a draft EEC Directive was prepared. This Directive was adopted by the Community in 1985 and became law in 1988. Although a Directive is a Community Law which is binding as to the results to be achieved, it is left to the member states as to the choice and methods of implementation.

The methods of implementation have been widely discussed and so too have been the two lists of projects; list one includes schedule 1 projects for which an EIA is required in every case, list two includes schedule 2 projects for which an EIA is required only if the particular project in question is judged likely to give rise to significant environmental effects. Schedule 1 projects include crude-oil refineries, major chemical and steel works, aerodromes with runways over 2100 m in length, and the permanent storage or final disposal of radioactive and toxic waste. Schedule 2 projects include large pig and poultry rearing units, mineral extraction, and metal processing, chemical, food, textile or rubber industries.

Environmental assessments, although not in the formal framework of EIAs, have long been undertaken in Britain in various forms for many years. Oil exploration in the North Sea prompted environmental assessments; for example, a formal EIA was undertaken in Britain as long ago as 1977 when SPHERE Environmental Consultants Ltd published an EIA report in several volumes for the consortium which developed the North Sea Beatrice oil field. During the period of about 1973 to 1988, there were nearly 150 environmental impact studies undertaken in relation to gas and oil exploration around Britain.

The European Council Directive (on the assessment of the effects of certain public and private projects on the environment) has 14 descriptive articles and three annexes; one lists projects for which EIAs will be mandatory, one lists projects for which EIAs may be carried out at the discretion of a member state, the third annex sets out the information required in an EIA. Although the word 'monitoring' does not appear, monitoring would be presumed because in the introduction it says

Whereas the effects of a project on the environment must be assessed in order to take account of concerns to protect human health, to contribute by means of a better environment to the quality of life, to ensure maintenance of the diversity of species and to maintain the reproductive capacity of the ecosystem as a basic resource for life;

and in Annex Three it says

5. A description of the measures envisaged to prevent, reduce and where possible offset any significant adverse effects on the environment.

Both of these statements would suggest that biological monitoring would be required although words such as 'significant' and 'maintenance' suggest that a precise or objective approach to monitoring would not necessarily be adopted.

The EIA process

A prerequisite for a good EIA is the preparation of a conceptual plan or framework. Conceptual frameworks have been developed as part of many EIAs, ranging from those EIAs designed for the Antarctic (Benninghoff & Bonner 1985) to those designed for impacts of electricity generating stations on the natural environment in temperate, coastal regions (Bamber 1989).

The precise steps for an EIA are dependent on legislation but in general there is a sequence of steps commencing with a project proposal and screening; a decision is taken as to whether or not an EIA should be undertaken. There can then be a consultation phase (Fig. 15.1). The consultation phase can be of considerable benefit to the developer, particularly with regard to the planning of the subsequent environmental assessments and in preparing a smooth passage for the whole environmental process. A project outline would usually incorporate information about the location and physical limits to the development. Whereas defining the physical limits of the site can be done objectively, defining the area affected by other impacts such as airborne effluents can not be undertaken in a definitive manner without detailed and prolonged studies and so may not be undertaken at all.

A preliminary outline EIA or scoping can be an important cost-effective stage where technical details of the development are prepared along with preliminary environmental information. The aim is to identify the scope and depth to which impact assessments should be undertaken and to focus attention on information gaps. Another important function of the preliminary EIA is to identify the expertise and personnel who will be eventually required to undertake the full environmental assessments.

Baseline data are the first step towards a monitoring programme. A description of the site could include information on physical, demographic

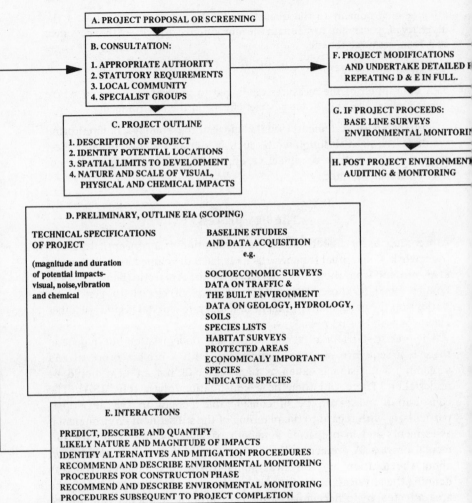

A. PROJECT PROPOSAL OR SCREENING

B. CONSULTATION:

1. APPROPRIATE AUTHORITY
2. STATUTORY REQUIREMENTS
3. LOCAL COMMUNITY
4. SPECIALIST GROUPS

F. PROJECT MODIFICATIONS
AND UNDERTAKE DETAILED F
REPEATING D & E IN FULL.

C. PROJECT OUTLINE

1. DESCRIPTION OF PROJECT
2. IDENTIFY POTENTIAL LOCATIONS
3. SPATIAL LIMITS TO DEVELOPMENT
4. NATURE AND SCALE OF VISUAL,
   PHYSICAL AND CHEMICAL IMPACTS

G. IF PROJECT PROCEEDS:
BASE LINE SURVEYS
ENVIRONMENTAL MONITORIN

H. POST PROJECT ENVIRONMEN
AUDITING & MONITORING

D. PRELIMINARY, OUTLINE EIA (SCOPING)

TECHNICAL SPECIFICATIONS
OF PROJECT

(magnitude and duration
of potential impacts-
visual, noise,vibration
and chemical

BASELINE STUDIES
AND DATA ACQUISITION
e.g.

SOCIOECONOMIC SURVEYS
DATA ON TRAFFIC &
THE BUILT ENVIRONMENT
DATA ON GEOLOGY, HYDROLOGY,
SOILS
SPECIES LISTS
HABITAT SURVEYS
PROTECTED AREAS
ECONOMICALY IMPORTANT
SPECIES
INDICATOR SPECIES

E. INTERACTIONS

PREDICT, DESCRIBE AND QUANTIFY
LIKELY NATURE AND MAGNITUDE OF IMPACTS
IDENTIFY ALTERNATIVES AND MITIGATION PROCEEDURES
RECOMMEND AND DESCRIBE ENVIRONMENTAL MONITORING
PROCEDURES FOR CONSTRUCTION PHASE
RECOMMEND AND DESCRIBE ENVIRONMENTAL MONITORING
PROCEDURES SUBSEQUENT TO PROJECT COMPLETION

Fig. 15.1. A simplified outline for an Environmental Impact Assessment showing where environmental monitoring could be included. Environmental monitoring includes chemical as well as biological and ecological monitoring.

and related infrastructure features. Areas of conservation or landscape interest would be identified but biological information is usually confined to protected species, sensitive species, taxa of commercial interest, taxa of scientific importance and taxa perceived to be important by the local community.

Potential impacts may be easy to identify but then comes the more difficult task of assessing the interactions between impacts and the various components of the environment. This part of the process is directed largely towards the human environment or those aspects of the environment which are of immediate concern to man's use of the environment. There are three basic kinds of methods for identifying impacts: flow diagrams or networks, matrices, and checklists. Of these, checklists are seemingly the most simple, but various methods have been refined to quantify the impacts. For example, the Battelle Environmental Evaluation Scheme (Dee *et al.* 1973), which is particularly comprehensive, was designed to assess impacts arising from water resources developments. The impacts are grouped into four main categories (ecology, environmental pollution, aesthetics, human interest) and these in turn are divided into components and parameters. There is an interesting next step in which each component of the checklist is transformed (see previous examples of normalization in Figs 8.1 and 12.2) into a level of environmental quality ranging from 0 (poor) to 1.0 (very good). Examples of these transformations for species diversity, dissolved oxygen and physical appearance of water are shown in Fig. 15.2. The principle seems good in that there is an attempt to quantify parameters of the environment and changes in these parameters can be quantified in terms of changing environmental quality. The relationship does, of course, require a subjective judgement as to what equates with different levels of environmental quality. There are, for example, some plant communities, such as heathlands, which characteristically have a low plant species diversity and any increase in diversity would indicate a diminished quality and not an increase in quality of the heathland community.

Other well-known methods for identification of impacts have been

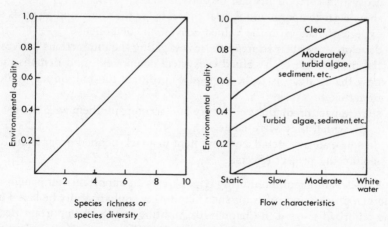

Fig. 15.2. Value function graphs employed in the Battelle Environmental Evaluation System. After Canter (1977).

based on matrices, some of which can be used to give a subjective assessment of the magnitude and importance of the impact. The Leopold Matrix (Fig. 15.3) is perhaps the most well-known matrix which has been used in EIAs (Leopold *et al.* 1971, Parker & Howard 1977). In that matrix, environmental characteristics and project actions are recorded in an open matrix with up to 8800 interactions. First, the matrix is used to record those impacts (interactions) which are likely to occur and if the impact is 'significant' then a diagonal line is entered in a box. The interaction is then scored from 1 to 10 (1, low, and 10, high) for both magnitude of the interaction (upper left) and the relative importance (lower right). Despite the criticisms that have been levelled at the Leopold Matrix (for example, it does not cater for secondary or tertiary interactions) this kind of approach has a cost-effective role to play in providing a basis for monitoring changes in ecological systems.

The biological and ecological components of EIAs leave a lot to be desired but the biological monitoring components have been virtually non-existent. The study by Beanlands & Duinker (1984) on ecological frameworks of EIAs was especially criticial of the sporadic applications of ecological principles in impact assessments. On the basis of their review of 30 Environmental Impact Statements prepared in Canada, they made a number of recommendations for new initiatives and also drew attention to monitoring as one of the EIA requirements. They suggested that an EIA should be required to:

1. identify early on, an initial set of valued ecosystem components to provide a focus for subsequent activities;
2. define a context within which the significance of changes in the valued ecosystem components can be determined;
3. show clear temporal and spatial contexts for the study and analysis of expected changes in the valued ecosystem components;
4. develop an explicit strategy for investigating the interactions between the project and each valued ecosystem component, and demonstrate how the strategy is to be used to co-ordinate the individual studies undertaken;
5. state impact predictions explicitly and accompany them with the basis upon which they were made;
6. demonstrate and detail a commitment to a well-defined programme for monitoring project effects.

Post-development monitoring (Fig. 15.1) is an important component of the EIA process and plans for such monitoring could usefully be based on the criteria discussed in Chapter 10. Auditing is also an important post-development component which is undertaken to test the scientific accuracy of the environmental impact predictions.

THE LEOPOLD MATRIX

REDUCED GRID SHOWING ALL LIKELY IMPACTS

EXISTING CONDITIONS OF THE ENVIRONMENT	PHYSICAL AND CHEMICAL CHARACTERISTICS																FLORA				FAUNA					CULTURAL FACTORS							ECOLOGY		
	Soils	Landform	Surface water	Underground water	Water quality	Water recharge	Atmospheric quality	Microclimate	Temperature	Floods	Erosion	Deposition	Solution	Compaction and settling	Stability	Air movements	Trees	Shrubs	Microflora	Barriers	Birds	Land animals	Insects	Microfauna	Barriers	Land use-wilderness	Scenic views	Oper. space qualities	Employment	Population density	Waste disposal	Man-made barriers	Food chains	Brush encroachment	
Modification of habitat	4/2	2/3	2/3		1/2			3/1								2/1	5/4	3/3	3/3	4/2	3/3	3/4	2/1	2/1	3/4	6/6	4/5	5/6					3/4	3/2	63/60
Modification of ground cover	3/2		1/2		1/2			2/1		3/5	3/4	2/4	1/1	3/3	3/4	2/1	5/4	3/3	2/3			4/4	2/1	2/1				6/6					2/3		50/54
Modification of ground-water	3/2		2/3	3/2	3/3	4/3				5/4			2/1		5/4																				27/22
Alteration of drainage	2/3		4/4	4/2	2/2	4/3				4/6		3/2			2/3																				25/25
Burning	3/3						3/1				2/2	1/2					5/6	5/4	4/4		4/5	5/5	3/3	4/3											39/38
Surface changes or paving	2/2	2/4	2/1		3/2									2/3			4/4	3/4	3/3	2/3		4/5	2/2	2/3				5/4							36/40
Noise and vibration							5/2							6/1	6/2						4/2	5/3	4/2	5/2											35/14
Urbanisation	3/2	4/2	3/3	2/1	2/4		4/4	3/2	2/1	3/5	2/3	2/3		1/1	1/2	1/1	4/5	3/4	3/4	2/3	4/5	5/5	3/3	3/3	4/4	5/6	6/6	6/6	4/2	5/4	4/4	3/1			97/99
Roads and trails	2/2	3/2			2/3		3/2	1/1		3/4	2/2	2/2		1/1	2/3	1/1	3/5	3/4	2/3	2/3	4/4	4/5	2/3	2/3	3/5		4/5	5/6	3/2					2/1	61/72
Transmission lines and pipelines		2/1		2/2				1/1						1/1	2/2	1/1											4/5		3/2						16/15
Barriers and fences		1/1						1/1			2/1					1/1						3/4		2/3			5/5	5/4				2/1			22/21
Cut and fill	3/3	4/4	2/3		3/3			2/1		3/3	2/3	2/3		2/2	3/2		3/4	2/3	2/3	2/3		4/5			3/5		3/6	4/6	3/2			2/1			54/65
Landscaping	1/1	4/2	3/2	2/2			1/1	1/1		3/4	2/3	2/3		1/2	2/3	1/1	3/5	2/4	2/4	3/2		5/4	2/2	2/2	4/5	5/6	5/6	4/4	3/3					1/1	64/73
Traffic by automobile					2/4		5/5	3/2									3/5	3/4	2/4		5/5	5/5	3/3	3/3											34/40
Traffic on trails	1/1						2/3										2/4	2/4	1/1		4/4	4/5	1/2	1/2				2/4						2/2	22/32
Spills and leaks	3/4			3/4	3/5					2/4					3/4																				14/21
	30/27	22/19	19/21	16/13	22/30	8/6	23/18	17/11	2/1	26/35	15/18	14/19	3/2	17/14	29/29	9/7	37/46	29/37	24/32	15/16	28/28	51/54	24/22	28/26	17/23	16/18	31/37	42/46	16/11	5/4	4/4	7/3	5/7	8/6	

Fig. 15.3. An example of a condensed version of the Leopold Matrix for an area of urban development near Southampton. From Spellerberg (1981).

There is a lack of good examples of monitoring as a part of the EIA process but there have been some attempts to incorporate baseline studies and ecological monitoring of the impacts which have or may have been caused as a result of oil and gas exploration and recovery. It is for this reason that we focus particularly on monitoring undertaken in relation to environmental impacts of oil spills.

Incidence of oil spills

The growth of petroleum installations and the impacts caused by pollution from petroleum installations and tankers around the United States and the British Isles have prompted a number of discussions about environmental impact assessment and monitoring programmes for terrestrial, freshwater and marine ecosystems. The Nature Conservancy Council has, for example, published guidelines for onshore oil and gas development which refer to environmental assessments and an environmental impact checklist (NCC 1986). Ecological monitoring, however, receives but a cursory mention in these guidelines without any indication as to how or why monitoring should be undertaken.

It is perhaps the oil spills, especially those polluting coastal and estuarine areas, which have received most attention and publicity, and indeed much research has been undertaken on monitoring the effects of oil spills on coastal and estuarine regions. This is not to suggest that an equal amount of research should be directed towards the prevention of oil spills and a rapid response to oil spills where ever they occur. Estimates of the levels of marine pollution from petroleum hydrocarbons have been researched by various organizations including the IUCN (1983) and NAS (1975). It seems generally agreed that about 3.5 million tonnes of petroleum hydrocarbons enter the world's marine environment each year, either directly or indirectly from all sources (Samiullah 1985). It is difficult to put this kind of figure into perspective and of more direct interest is the number of oil spills which have occurred.

Since 1971 the US Coast Guard has been collating data on oil spills and during 1974, for example, there were 11 440 reported oil pollution incidents of which vessels accounted for 26 per cent (Gundlach & Hayes 1978). No coastline (including the remote coast of Antarctica – see Chapter 1) is safe from the potential threat of oil spills. In 1988, for example, an Ecuadorian supply ship hit a reef off the Galapagos Islands and 50 000 gallons of diesel fuel ended up on Santa Cruz. There are as many as 500 oil spills around Britain each year. There were at least 11 major incidents during 1988 and 1989 (Fig. 15.4).

From a biological monitoring point of view, we can recognize five

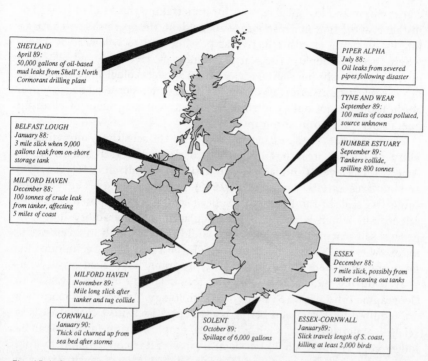

Fig. 15.4. Some of the major oil spills which occurred around Britain in 1988–89. In some years there may be as many as 500 oil spills.

responses to these oil spill incidents: 1, research on methods of detecting, recovering and clean-up operations; 2, research on the effects of oil on ecosystems; 3, baseline surveys for future monitoring; 4, ecological monitoring to ensure effective management; and, 5, monitoring after an oil spill incident to follow the progress of ecological recovery. The first two of these have been much researched but are well outside the scope of this book. The latter three responses are of direct interest to monitoring and although not all examples described here have been integral parts of formal EIAS, there has been a growth in EIAS incorporating baseline surveys and environmental assessments as well as ecological monitoring.

Baseline surveys and environmental assessments

Some of the research relevant to ecological baseline surveys can be found in the interesting methods previously used for classification of the level of sensitivity and vulnerability of coastlines to oil spills. There are good conservation and logistical reasons for undertaking this kind of work. For instance, at the time of a response to an incident, the operators will have an interest in being as cost-effective as possible and so will need to know where best to deploy the various kinds of recovery and clean-up operations. Classifying coastal regions from a strict ecological point of view or via subjective assessments of sensitivity to water-borne oil has practical applications in not only contributing to better management of clean-up operations but also in siting installations.

The techniques for assessing and recording the sensitivity and vulnera-bility of coastal areas has been particularly well researched in North America by Hayes, Gundlach & Getter (1980), Gundlach & Hayes (1978) and more recently by Jensen et al. (1990). Gundlach & Hayes developed a subjective scale of 1 to 10 based on type of coast (rocky, exposed headland, salt marsh) which is the basis for calculating a vulnerability index. In general salt marshes and mangroves would be rated high in terms of the index whereas exposed headlands would be rated low. Techniques for Environmental Sensitivity Index (ESI) mapping have been greatly im-proved during the last few years and this is largely due to advances in Geographic Information System (GIS) technology. As more and more data are required for better and better assessments of sensitive coastal areas in relation to perturbations such as oil spills, simple distribution maps can no longer cope with the amount of data. However, it is now possible to store information about the spatial and temporal distribution of sensitive taxa in GIS (see p. 174) which not only allows rapid retrieval but also quite complex manipulation of the data (Jensen et al. 1990).

Environmental assessments of coastlines in relation to the sensitivity of habitats and regions to water-borne oil have also been undertaken in relation to anti-pollution strategies at Sullom Voe in the Shetlands (Syratt & Richardson 1981) and also in the area around Poole Harbour, on the south coast of England (Pearson 1984, Gray 1985, B.P. Petroleum Development 1986) where BP have undertaken a number of EIAS. Each section of the shoreline in Poole Harbour was assessed in terms of its relative sensitivity to damage by water-borne oil. Intertidal plants and some animals, particularly birds, provided the basis for the preparation of a composite shoreline 'ecological' sensitivity map. The limitations of this sensitivity assessment (as discussed in the publication) include the following: the assessment does not take full account of the interde-pendence of different parts of the harbour ecosystem, intertidal ecosy-stems are dynamic and therefore such an analysis is essentially a 'snap-shot' of present conditions, and assessment of ecological sensitivity specifically ignores questions of land use or evaluation. However, despite these and other limitations, sensitivity maps of this kind form the basis for oil spill contingency plans and a possible basis for monitoring changes in the area which has been assessed.

There remains much disagreement about how long different ecosystems take to recover and there is also disagreement about the criteria used to establish when an area has recovered from the effects of an oil spill. The criteria have ranged from the renewed occurrence of organisms (no matter what they may be) to balanced age-class structure in the coastal species. At high latitudes, many species may be at the limits of their distribution and recolonization after an oil spill may be a remote possibility. In addition, many marine species do not show regular recruitment and therefore it could be argued that it is unrealistic to define recovery when a balanced age-class has been reached (Clark 1982, 1989). Underlying the disagreements about time-scales and criteria is a lack of research, especially on those biotic communities outside temperate climatic regions. Polar communities and some diverse and complex tropical communities, especially mangroves and coral reefs, require much more research so as to be able to judge when a damaged ecosystem has recovered.

Monitoring coastal communities

The CONCAWE Group (oil companies international study group for conservation of clear air and water – Europe) was established in 1963 and they have summarized some of the ecological monitoring procedures (including ideas for sampling techniques and data analysis) for aqueous effluents from petroleum refineries (CONCAWE 1982). Ideas for data analysis incorporate the use of biotic indices, diversity indices, similarity indices and simple diagrammatic presentations showing relative abundance. Obviously there is as much choice of methods and analysis as there is variation in coastal communities and conditions. The variation in coastal communities means therefore that different monitoring methods are required for the different communities. Generally speaking, data collection can be of two kinds: removal of material or destructive sampling, and non-destructive recording techniques. The latter includes satellite imagery, aerial photography, stereophotography, and visual recording.

One other interesting consideration and one which tends to be forgotten until faced with samples of material is the identification of specimens. There are for example many coastal areas, such as on some of the coasts of Alaska (Cowell & Monk 1981), where the fauna and flora have not been studied in detail. Taxonomic keys may not be available for some groups and some groups such as marine algae may include many species which are simply difficult to identify. Taxonomic difficulties highlight the dependence of biological and ecological monitoring on taxonomy and it is regrettable that there seem to be fewer and fewer taxonomists graduating from institutions of higher education.

To illustrate some of the applications of methods used for monitoring ecological effects of oil spills, we can usefully refer briefly to a selection of case studies based on different types of communities such as rocky shores, salt marshes, mudflats and subtidal communities (rocky and soft sediments).

Rocky shores

The monitoring of the biological effects of the development of the oil terminal at Sullom Voe in the Shetlands is one example of a carefully designed part of a large EIA (Dicks 1989). The biological monitoring commenced in 1978 and has been a large integrated programme directed at a range of taxonomic groups and habitats including rocky shores. At each site selected for surveying and later monitoring, permanent markers identified the sampling area (see comments about permanent markers on p. 88). In this instance the permanent markers were paint spots and many photographs were also taken to ensure that the sampling site could readily be located. The data were collected (during spring tides) at 20 cm intervals along transects from low water up to the lower limit of the flowering plants. Abundance of lichens, algae and animals were recorded within a 3 m wide strip. The abundance scale used by Hiscock (1981) was similar to that shown in Table 15.1 and was modified only slightly, in respect to specific organisms, from previously well-tried methods of abundance estimation for rocky shores (Crapp 1973). Photographs were also used to supplement the visual inspections of the monitoring sites. Similar abundance scales were used in ecological monitoring in Port Valdez, Alaska (Cowell & Monk 1981) in relation to the export of Prudhoe Bay crude oil. These abundance scales have also been used, since 1972, to monitor rocky shore organisms in the vicinity of a refinery at Mongstad in Norway (CONCAWE 1982, Dalby, Cowell & Syratt 1979). In addition size-class distributions of the mollusc *Patella vulgata* were monitored at some sites to investigate levels of population recruitment.

It is perhaps important to note that the levels of monitoring described above can detect only gross changes in the ecosystems and communities but nevertheless this rocky shore monitoring, as an integral component of a much larger biological monitoring programme, has provided important data from which to assess the impacts during construction and even during extreme perturbations caused by oil spill clean-up operations (Westwood, Dunnet & Hiscock 1989, CONCAWE 1982).

Table 15.1. *Abundance values for ecological monitoring of rocky shores*

1. Flowering plants and lichens
   Ex.   More than 80% cover
   S.   50–80% cover
   A.   20–50% cover
   C.   1–20% cover
   F.   Large scattered patches
   O.   Widely scattered patches, all small
   R.   Only one or two patches

2. Seaweeds
   Ex.   More than 90% cover
   S.   60–90% cover
   A.   30–60% cover
   C.   5–30% cover
   F.   Less than 5% cover, zone still apparent
   O.   Scattered plants, zone indistinct
   R.   Only one or two plants

3. Barnacles (excluding *B. perforatus*) *Littorina neritoides*, small forms of *L. saxatilis*
   Ex.   More than 4 per sq. cm.
   S.   3–5 per sq. cm.
   A.   1–3 per sq. cm.
   C.   10–100 per sq. decimetre
   F.   1–10 per sq. decimetre, never more than 10 cm. apart
   O.   1–100 per sq. metre, few within 10 cm. of each other
   R.   Less than 1 per sq. metre

4. *Balanus perforatus*
   Ex.   More than 3 per sq. cm.
   S.   1–3 per sq. cm.
   A.   10–100 per sq. decimetre
   C.   1–100 per sq. metre
   F.   10–100 per sq. metre
   O.   1–10 per sq. metre
   R.   Less than 1 per sq. metre

5. Limpets and Winkles (excluding *Littorina neritoides* and small forms of *L. saxatilis*)
   Ex.   More than 200 per sq. metre
   S.   100–200 sq. metre
   A.   50–100 per sq. metre
   C.   10–50 per sq. metre
   F.   1–10 per sq. metre
   O.   1–10 per sq. decametre
   R.   Less than 1 per sq. decametre

6. Gastropods (excluding Limpets and Winkles)
   Ex.   More than 100 per sq. metre
   S.   50–100 per sq. metre
   A.   10–50 per sq. metre
   C.   1–10 per sq. metre, locally sometimes more
   F.   Less than 1 per sq. metre, locally sometimes more
   O.   Always less than 1 per sq. metre
   R.   Less than 1 per sq. decametre

Table 15.1. (*cont.*)

7. Mussels
   Ex.   More than 80% cover
   S.   50–60% cover
   A.   20–50% cover
   C.   Large patches, but less than 20% cover
   F.   Many scattered individuals and small patches
   O.   Scattered individuals, no patches
   R.   Less than 1 per sq. metre

8. *Pomatoceros triqueter*
   A.   More than 50 tubes per sq. decimetre
   C.   1–50 tubes per sq. decimetre
   F.   10–100 tubes per sq. metre
   O.   1–10 tubes per sq. metre
   R.   Less than 1 tube per sq. metre

9. *Spirorbis* spp.
   A.   5 or more per sq. cm.: on 50% of suitable surfaces
   C.   5 or more per sq. cm.: on 5–50% of suitable surfaces
   F.   1–5 per sq. cm.: on 1–5% of suitable surfaces
   O.   Less than 1 per sq. cm.
   R.   Less than 1 per sq. metre

From CONCAWE (1982) after Crisp & Southwood (1958) and Crapp (1973).

Salt marshes

The monitoring which has been undertaken on the *Spartina* salt marshes adjacent to the Fawley refinery on Southampton Water has already been described in detail (see p. 170) as an example of ecosystem monitoring.

Mudflats

Since 1975, ecologists at Stirling University have monitored the effects of aqueous effluents on intertidal soft sediment communities around the oil refinery at Grangemouth (east coast of Scotland). There, the monitoring is based on mud samples taken at points along 18 permanent transects (McLusky, Elliott & Warns 1978, Cowell 1978) and abundance and distribution of sentinel indicator species (see Chapter 5), such as the mollusc *Hydrobia ulvae* are recorded. Based on the species composition, distribution and abundance of various organisms, it has been possible to make subjective assessments of 'pollution zones' of the mudflats which then form the basis of pollution zone maps. These maps have provided a basis to assess the quality and extent of the management operations and the assessment has resulted in improved management operations.

Table 15.2. *The benthic macrofauna at a sampling site near Naantali Harbour (1965–76)*

	1965	1967	1970	1972	1973	1974	1975	1976
No. of species	4	4	3	4	5	9	9	9
Density (ind/m²)	250	530	400	70	1650	1470	2634	1970
Biomass (g/m²)	6.5	7.9	17.9	0.4	27.6	29.6	157.4	282.5
Diversity	1.6	1.3	1.2	1.2	0.9	0.7	0.9	n.a.
Species richness	0.5	0.5	0.5	0.5	0.4	0.7	1.0	n.a.
Evenness	0.8	0.7	0.6	0.8	0.4	0.3	0.3	n.a.

n.a. = not available.
From CONCAWE (1982).

At Naantali in Finland, a more detailed ecological monitoring programme than described above has been established as a basis for
monitoring what were severe impacts from a refinery on the organisms of
sub-tidal soft sediments. Poor management of refinery effluents over a
period of about 15 years until 1972 had quite severe visible effects on
adjacent, coastal fauna. Following the commencement of an ecological
monitoring programme in 1970, there was an introduction of a new
treatment process in 1973 for the effluents. The sampling, undertaken by
staff from the University of Turku, took place within permanent
monitoring sites located by fixing on inshore landmarks. Sampling was
undertaken at control sites as well as adjacent to the outflow from the
refinery.

The usual abundance records of selected taxa have been recorded along
with measurements of biomass. Species richness and species diversity of
benthic littoral macro-organisms were also recorded both before and after
the new effluent treatment had been introduced. Results from one of the
inshore sites shows the effect of controlling and treating the effluent (Table
15.2).

Research and development

Apart from any technical and logistic problems of monitoring, the spatial
and temporal changes in communities pose one of the greatest problems for
ecological monitoring. In other words, if the organisms were distributed
evenly on the shores and did not change over time, then monitoring would
be made much easier. Monitoring change over time requires a decision
about frequency of sampling: too little sampling may not reveal seasonal or
shorter patterns and too much sampling may damage (via physical
disturbance) the habitat. Spatial patterns are no easier to detect. As well as

being accurate, the sampling area has to be adequate and representative. On the basis of rocky shore monitoring around the Isle of Man (UK), Hartnoll & Hawkins (1980) suggested that where resources are limited, the concentration of effort by sampling larger areas could be offset by sampling at a reduced number of stations. Clearly, long-term ecological studies could provide data of much value to the design of effective coastal monitoring programmes.

We have seen that much attention has been given to methods of scoring abundance but there has obviously been less attention to research that may be required for establishing natural spatial and temporal changes, especially long-term changes in coastal communities. This has been noted in relation to long-term ecological studies still continuing today by ecologists of the BP Company at the Rafinor oil refinery at Mongstad on Fernsfjord in Norway. It has become very clear after many years of monitoring that change in biological communities is the rule rather than the exception. The main problem there has not been so much detecting change but assessing its significance in attributing a cause (Dalby et al. 1979).

New methods of ecological monitoring of the effects of pollution are always being developed and one good example has been the research by Christie (1980) on rocky subtidal communities in Norway. The aim of the research was to combine the use of stereophotography, field experiments and manipulation as a method for monitoring effects of chronic pollution. Two experiments and a control were established in a site of about 2.0 m². The aim of one experiment was to record the recolonization of a small cleared area and the second experiment used exclusion cages so as to examine the effects of predation. Photographs of the percentage cover of different organisms was the main source of data.

After 14 months, the results of these experiments were showing that predation by *Coryphella*, *Asterias* and *Psammechinus* was the most important factor affecting the structure of the community. Christie concluded that these predators are 'key' species and therefore an effective monitoring programme would need to record the abundance and effects of these predators on the sub-tidal community.

References

Bamber, R. N. (1989). *Environmental Impact Assessment: the Example of* CEGB *Power Stations and Marine Biology*. Report RD/L/3524/R89. Leatherhead, UK, Central Electricity Board.
Beanlands, G. E. & Duinker, P. N. (1984). An ecological framework for environmental impact assessment. *Journal of Environmental Management*, **18**, 267–77.

Benninghoff, W. S. & Bonner, W. N. (1985). *Man's Impact on the Antarctic Environment: A Procedure for Evaluating Impacts from Scientific and Logistic Activities.* Cambridge, SCAR and the Scott-Polar Institute.

BP Petroleum Development Ltd. (1986). *Wytch Farm Development, Purbeck–Southampton Pipeline Environmental Impact Assessment.* BP Petroleum Development Ltd.

Canter, L. W. (1977). *Environmental Impact Assessment.* London, New York, McGraw-Hill.

Christie, H. (1980). Methods for ecological monitoring: biological interactions in a rocky subtidal community. *Helgolander Meeresuntersuchungen*, 33, 473–83.

Clark, M. & Hetherington, J. (1988). *The Role of Environmental Impact Assessment in the Planning Process.* London, Mansell.

Clark, R. B. (1982). *The Long-term Effects of Oil Pollution on Marine Populations, Communities and Ecosystems.* London, The Royal Society.

Clarke, R. B. (1989). *Marine Pollution.* Oxford, Clarendon Press.

CONCAWE (1982). *Ecological Monitoring of Aqueous Effluents from Petroleum Refineries.* Report no. 8/82, CONCAWE, Den Haag.

Cowell, E. B. (1978). Ecological monitoring as a management tool in industry. *Ocean Management*, 4, 273–85.

Cowell, E. B. & Monk, D. C. (1981). Problems in ecological monitoring in Port Valdez, Alaska. *Proceedings of the Royal Society of Edinburgh*, 80B, 355–66.

Crapp, G. B. (1973). The distribution and abundance of animals and plants on the rocky shore of Bantry Bay. *Irish Fisheries Investigations*, series B, No. 9.

Crisp, D. J. & Southwood, A. J. (1958). The distribution of intertidal organisms along the coasts of the English Channel. *Journal of the Marine Biological Association*, 37, 157–208.

Dalby, D. H., Cowell, E. B. & Syratt, W. J. (1979). Biological monitoring around an oil refinery. In *Monitoring the Marine Environment*, ed. D. Nichols, pp. 141–52. Symposia of the Institute of Biology No. 24. London, IOB.

Dee, N., Baker, J. K., Drobny, N. L., Duke, K. M., Whitman, I. & Fahringer, D. (1973). Environmental evaluation system for water resource planing. *Water Resources Research*, 9, 523–35.

Dicks, B. (1989). *Ecological Impacts of the Oil Industry*, Proceedings of an International Meeting Organized by the Institute of Petroleum and held in London in November 1987. Chichester, New York, John Wiley, on behalf of the Institute of Petroleum.

Gray, A. J. (1985). *Poole Harbour: Ecological Sensitivity Analysis of the Shoreline.* Huntingdon, NERC Institute of Terrestrial Ecology.

Gundlach, E. R. & Hayes, M. O. (1978). Vulnerability of coastal environments to oil spill impacts. *Marine Technology Society Journal*, 12, 18–27.

Hartnoll, R. G. & Hawkins, S. J. (1980). Monitoring rocky-shore communities: a critical look at spatial and temporal variation. *Helgolander Meeresuntersuchungen*, 33, 484–94.

Haug, P. T., Burwell, R. W., Stein, A. & Bandurski, B. L. (1984). Determining the significance of environmental issues under the national Environmental policy Act. *Journal of Environmental Management*, 18, 15–24.

Hayes, M. O., Gundlach, E. R. & Getter, C. D. (1980). Sensitivity ranking of energy port shorelines. In *Proceedings of Ports*, ed. M. O. Hoyes et al.), pp. 697–709. Norfolk, Virginia, American Society of Civil Engineers.

Hiscock, K. (1981). The rocky shore ecology of Sullome Voe. *Proceedings of the Royal Society of Edinburgh*, 80B, 219–40.

IUCN (1983). *Impact of Oil Pollution on Living Resources*. J. M. Baker (Chairman), Working group on oil pollution of the IUCN Commission on Ecology in co-operation with the WWF. Commission on Ecology papers, No. 4. International Union for the Conservation of Nature and Natural resources.

Jensen, J. R., Ramsey, E. W., Holmes, J. M., Michael, J. E., Savitsky, B. & Davis, B. A. (1990). Environmental sensitivity index (ESI) mapping for oil spills using remote sensing and geographical information system technology. *International Journal of Geographical Information Systems*, 4, 181–201.

Leopold, L. B., Clarke, F. E., Henshaw, B. B. & Balsley, J. R. (1971). A procedure for evaluating environmental impact. *Circular, U.S. Geological Survey*, p.645.

McLusky, D. S., Elliott, M. & Warns, J. (1978). *The Impact of Pollution on the Intertidal Fauna of the Estuarine Firth of Forth. Physiology and Behaviour of Marine Organisms*. Proceedings of the 12th European Symposium on Marine Biology. Oxford, Pergamon.

NAS (1975). *Petroleum in the Marine Environment*, E. B. Wilson (Chairman), report of a workshop on inputs, fates and effects of petroleum in the marine environment, Airlie, VA, 21–25 May 1973. Washington, National Academy of Sciences.

NCC (1986). *Nature Conservation Guidelines for Onshore Oil and Gas Development*. Peterborough, NCC.

Parker, B. C. & Howard, R. V. (1977). The first environmental impact monitoring and assessment in Antarctica. The Dry Valley drilling project. *Biological Conservation*, 12, 163–77.

Pearson, N. Associates (1984). *Wytch Farm Oilfield Development, Furzey Island Visual Impact Analysis*. BP Development Ltd.

Roberts, R. D. & Roberts, T. M. (1984). *Planning and Ecology*. London, New York, Chapman & Hall.

Rosenberg, D. M., Resh, V. H., Balling, S. S., Barnby, M. A., Collins, J. N., Durbin, D. V., Flynn, T. S., Hart, D. D., Lamberti, G. A., McElravy, E. P., Wood, J. R., Blank, T. E., Schultz, D. M., Marrin, D. L. & Price, D. G. (1981). Recent trends in environmental impact assessment. *Canadian Journal of Fisheries and Aquatic Sciences*, 38, 591–624.

Samiullah, Y. (1985). Biological effects of marine oil pollution. *Oil & Petrochemical Pollution*, 2, 235–64.

Spellerberg, I. F. (1981). *Ecological Evaluation for Conservation*. London, Edward Arnold.

Syratt, W. J. & Richardson, M. G. (1981). Anti-oil pollution in Sullom Voe – environmental considerations. *Proceedings of the Royal Society of Edinburgh*, 80B, 35–51.

Wathern, P. (1988). *Environmental Impact Assessment*. London, Unwin Hyman.

Westman, W. E. (1985). *Ecology, Impact Assessment, and Environmental Planning*. New York, Chichester, John Wiley.

Westwood, S. S. C., Dunnet, G. M. & Hiscock, K. (1989). Monitoring the Sullome Voe Terminal. In *Ecological Impacts of the Oil Industry*, ed. B. Dicks, pp. 261–85. Chichester, New York, John Wiley, on behalf of the Institute of Petroleum.

Wolfe, D. A. (1988). Urban wastes in coastal waters: assimilative capacity and management. In *Oceanic Processes in Marine Pollution*. Vol. 5, *Urban Wastes in Coastal Marine Environments*, ed. D. A. Wolfe & T. P. O'Connor, pp. 1–20. Malabar, Florida, Robert E. Kreiger Publishers.

16
Species monitoring and conservation

Introduction

SPECIES MONITORING is about more than just monitoring how many species are endangered and monitoring the rate at which they are becoming extinct. The populations of various species have been monitored for a number of reasons but the most common objectives include the following: to obtain data as a basis for sustainable use of populations; to provide a basis for detecting effects of environmental impacts; to provide data as a scientific basis for conservation. The last of these objectives is the main theme of this chapter. At the same time the value of long-term ecological studies is emphasized.

Objectives of species monitoring programmes

Monitoring species as a basis for sustainable use

Between 1950 and 1969 the world's fisheries catch rose steadily at about 6–7 per cent annually from 20 million to 65 million tons (WCED 1987). The fish stocks became more and more depleted and after 1970 the average annual growth in catches fell to only about 1 per cent. It is estimated that the world's catch is now 24 per cent less than it might have been with effective management of fish stocks. If ever there were a good example to justify the need for biological monitoring as a basis for calculating sustainable use of a natural resource, then overfishing of the world's oceans could not be a better example. In broad and simple terms, overfishing is a world problem and although not entirely a biological problem (politics and economics are part of the problem), monitoring is one of the key elements to solving that problem.

The methods used for monitoring fish stocks are particularly interesting and here we can briefly summarize some aspects researched by the Marine Laboratory staff at the Department of Agriculture and Fisheries (DAFS) in Scotland (P. J. Hopkins pers. comm.). Most fish stocks are assessed using a simple analysis called the Virtual Population Analysis (VPA) which is based on total mortality rate (natural mortality plus fishing mortality) and the proportion of fish surviving from one year to the next. The VPA is unreliable for the most recent years, partly because the natural mortality rate is usually guessed, but surveys may be used to estimate stock sizes in any current year. The survey methods include fishing surveys, acoustic surveys and egg and larvae surveys.

Fishing surveys can be used to obtain abundance indices of various fish species, particularly the juveniles which have not been taken by the fishing industry. Estimating juvenile abundance is vitally important for projections of stock size, especially where there is much dependence on population recruitment. In some cases the survey area is divided into statistical rectangles and one haul is taken from each rectangle. International cooperation is one important key to success in these fishing surveys. An index of abundance for each major fish species is then calculated using the number of fish caught per hour of trawling. Such a survey method tends to give most reliable information for demersal species (living near or at the bottom), though useful information on pelagic stocks has also been obtained from bottom trawl surveys. Variable weather conditions, the heterogeneous distribution of the fish, and the trawling methods do not help to make fish surveys reliable and there is some evidence that catchability has changed on certain surveys (Armstrong & Cook 1986).

A recent development has been the use of acoustic surveys of pelagic fish stocks and basically the strength of the reflected signal is used to estimate the biomass of the shoal of fish. To do this, the acoustic target strength of the fish must be known and so experiments with caged and wild fish have to be undertaken. Although research is progressing towards the stage of being able to identify fish species, no reliable method yet exists despite the fact that some species do produce echo-traces with a singular characteristic appearance. Problems encountered with this method include mixtures of species and the difficulty of detecting fish near the surface or at the bottom.

Egg and larvae surveys have long been used as a basis for monitoring fish stocks but not all species are suitable for this method. Not only must the spawning season be known but it must also be reasonably consistent from year to year. When the distribution of the spawning has been established, a decision then needs to be taken whether or not to estimate the absolute stock size or provide an index of the stock size. In waters around Britain, egg surveys have been used to assess Mackerel (*Scomber scomberus*) stocks and larvae surveys have been used to obtain indices of Herring (*Clupea harengus*) stocks.

One other classic example of a desperate need for species monitoring, although now too late, is the monitoring of whale populations. Ever since the establishment of the International Whaling Commission (p. 41), monitoring the status of whale populations has been crude, relying largely on material and data from commercial whaling operations. Commercial exploitation has supported the argument that whale catches should continue at such a level until someone else can show that stocks are diminishing. In the face of commercial operations it seems impossible to accept the approach that the biology of the species needs to be understood and that those exploiting the species should first show that levels of catch will not deplete the populations. It is ironic perhaps that despite severe criticism of the methods used to estimate whale populations and despite the criticism that those estimates have been too low, more recent research shows that Blue Whale (*Balaenoptera musculus*) populations are far lower than previously thought. It was announced at the 1989 meeting of the International Whaling Commission that there may be as few as 500 Blue Whales remaining in the southern oceans. In the Antarctic, whales have been subjected to so much exploitation that, despite total protection since 1967, there are few signs that any recovery has taken place (Bonner 1986).

Monitoring species as a basis for detecting effects of environmental impacts

Monitoring the occurrence of and the population ecology of a species has long been a basis for assessing the effects of environmental impacts. Baseline information is a prerequisite for this kind of biological monitoring programme and, because of natural populations changes, long-term monitoring in relation to the life cycle of a species is very necessary.

This kind of species monitoring has been undertaken in British waters by the Marine Laboratory at the Fawley Power Station. For example, 15 years' data have been collected on population dynamics of the Sand Smelt (*Atherina presbyter*) in Southampton Water (Henderson, Turnpenny & Bamber 1984), and 10 years' data have been accumulated on the population biology of the Common Shrimp (*Crangon crangon*) in the Severn Estuary and the Bristol Channel (Henderson & Holmes 1987). Fish movements as well as population dynamics have also been studied for many years; for example 10 years' data have now been obtained for fish movements in the Severn Estuary and the Bristol Channel (Henderson & Holmes 1989).

Monitoring species as a basis for conservation

For some species, the need for conservation is obvious. The numbers of
some species have become so low that, without action, extinction is
inevitable. In other cases, the fragmentation and isolation of many
populations may be the evidence required to show that conservation is
necessary. But conservation of wildlife is forced to be cost-effective and
with so many species becoming endangered or close to extinction, there is a
need to monitor the status of as many species as possible. Monitoring the
status of some species, particularly large furry and feathered species, has
been undertaken for decades but the species being monitored have been
chosen in an *ad hoc* manner. In the last two decades there has been a more
systematic approach to species monitoring for conservation with the
establishment of species monitoring programmes but even these programmes have not been without their problems (see Chapter 2).

Long-term population studies

Two examples of long-term animal population studies have already been
mentioned: the population studies of Herons initiated by Lack in 1928 and
which is continued today by the B T O (see Chapter 11). The other example
is the Rothamsted Insect Survey which continues to undertake long-term
studies on many species of insects (see Chapter 2 and Taylor 1986). Another
example of classic long-term insect population studies is the study of forest
insects established by Germany in the early 1800s (Klimetzek 1990), one
example of which is shown in Fig. 16.1. Other examples of long-term
studies on animal populations include work on Newts (Gill, Berven &
Mock 1983), Trout (Elliott 1984, 1988), Red Grouse (Moss & Watson 1985)
and Wolf and Moose populations in Michigan (Peterson 1987).

There have been few long-term plant population studies but some good
examples include grassland studies (Risser *et al.* 1981), forest studies (Peet
& Christensen 1987) and experiments at Rothamsted (see Chapter 3). A
now-classic example of long-term studies is the Breckland Plots (East
Anglia, England) established by Watt in 1936 (Davy & Jefferies 1981).
Examples of tenacious, long-term single-species studies include that
undertaken by Hutchings (1987) on the Early Spider Orchid (*Ophreys
sphegodes*) as shown in Fig. 16.2. Tamm (1972) and Wells (1981) have also
described long-term studies of orchid populations, and long-term studies
on *Primula scotica* have been published by Bullard *et al.* (1987). A database
of all North American long-term vegetation studies utilizing permanent
plots has been compiled at the Institute of Ecosystem Studies, New York
Botanical Garden (see Appendix II for address).

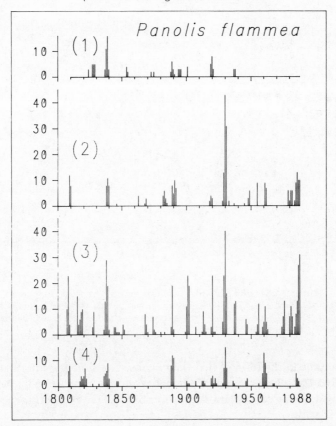

Fig. 16.1. Outbreaks of the Pine Beauty Moth (*Panolis flammea*) from 1801 to 1988 in the German forests of Pfalz (1), Oberpfalz (2), Mittelfranken (3) and Oberfranken (4). From Klimetzek (1990) with kind permission.

Monitoring for conservation

IUCN Red Data Books

In 1966, the IUCN Survival Service Commission was established to undertake a programme of action to prevent the extinction of plant and animal species and to ensure that viable populations were maintained in natural habitats. In order to achieve these ambitious aims, information of many species was required but that information was sometimes scattered throughout the world in many reports, scientific journals and books. The concept of a new system for bringing together information on various

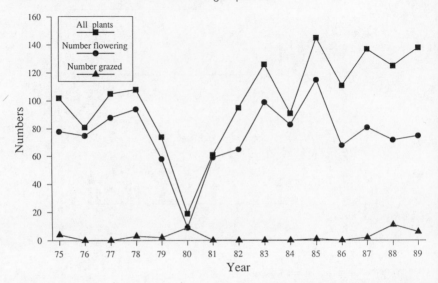

Fig. 16.2. Number of plants of *Ophrys sphegodes* observed (■), number flowering
(●) and number grazed (▲) at a site in Sussex, England. This work has been undertaken
by Hutchings (1987) and he has kindly allowed me to use data from 1984 onwards
which he has not yet published.

species came in the form of Red Data Books. The synoptic information in
these Red Data Books, culled from many sources, represented a basis for
not only devising conservation measures but also for monitoring the status
of the species.

National Red Data Books

Several countries, including Britain (Perring & Farrell 1983), Russia
(Borodin 1978), and New Zealand (Williams & Given 1981) have produced
their own Red Data Books as a basis for monitoring and assessing the
conservation needs of a species. In addition to Red Data Books, there are
many lists of threatened and endangered plants which have also been
produced by several countries including Africa (Hall *et al.* 1980), Australia
(Leigh & Boden 1979), Europe (Council of Europe 1977), North America
(Fairbrothers & Hough 1975, Kershaw & Morton 1976) and Russia
(Malyshev & Soboleuska 1980).

Britain's Red Data Book on vascular plants includes numerical information on plant species abundance as part of the assessment of threatened
status (there is a similar volume for Ireland, Curtis & McGough 1988). The
information in these Red Data Books is based on the change in the presence
of a species in 10 km² (see Fig. 3.1) over a certain period of time. The rate of

Table 16.1. *Suggested frequency of monitoring data for rare plant monitoring. (a) From Given (1983) for flora of New Zealand; (b) from Bowles et al. (1986) for flora at the Indiana Dunes National Lakeshore*

(a)	Frequency of monitoring	
Field	More often	Less often
Distribution	Few	Widespread
Population size	Small	Large
Habitat type	Seral	Climax
Life form, species type	r- (annual, etc.)	K- (long-lived, perennial, etc.)
Population structure	Imbalanced	Balanced (good percentage of age classes)
Reproduction	Poor	Good
Breeding system	Self-incompatible Dioecious	Self-compatible Monoecious
Pollinating vector	Specialized	Unspecialized
Threats	Actual, greater	Potential, lesser
Protected areas	None or small	Numerous
Cultivation	None or little	Widespread
Management	No deliberate programme	A deliberate programme
Taxonomic distinctiveness	Greater (endemic)	Less (subspecies, variety)

(b) Plant population type	Frequency of monitoring
Annual, biennial or rare	1–2 year intervals for reconnaissance monitoring
Perennial herbs, grasses	2–3 year intervals for reconnaissance monitoring
Dominant herbs, grasses	2–4 year intervals for reconnaissance monitoring
Trees or shrubs	4–5 year intervals for reconnaissance monitoring
Community dominants	5–10 year intervals for quantitative monitoring
Widespread plants not under threat	3–4 year intervals for quantitative monitoring
Rare plants or irregularly flowering plants	2–3 year intervals for quantitative monitoring
Successional	2–3 year intervals for quantitative monitoring

change in abundance can then be quantified and used along with other information, such as number of locations in protected areas, as a basis for calculating the threatened status of that particular species.

In New Zealand, Given (1983, 1988, 1989) has undertaken much research on the botany of that country and has put forward a convincing argument for the need to monitor the many threatened plant species and habitats. In his review of the application of monitoring techniques, Given (1983) addressed the need to be cost-effective and suggested a useful basis for the frequency of monitoring threatened plant species in New Zealand (Table 16.1). The frequency of monitoring depends in part on the biology of the

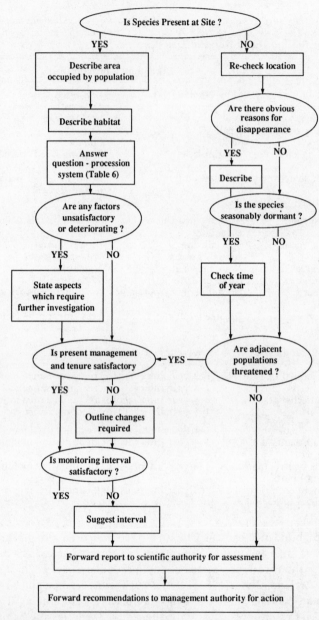

Fig. 16.3. Given's monitoring flow diagram for plant conservation in New Zealand. From Given (1983).

species being monitored and therefore any strategy for rare plant monitoring needs to be structured and designed accordingly. A flow chart (Fig. 16.3) showing operations for such a possible monitoring scheme has been suggested by Given (1983) and currently this is now being tested in New Zealand.

Rare plant monitoring in the USA

The number of plant monitoring projects to be initiated in the USA since 1974 has increased in a dramatic fashion (Fig. 16.4) and a timely assessment of the efficiency and accuracy of 172 rare plant monitoring projects was undertaken by Palmer in 1984 (Palmer 1986, 1987). The assessment (which basically asked the questions, what kind of monitoring is being done, by whom and where?) was initiated to promote communication among people involved in rare plant monitoring and to present a summary of the 'state of the art' in current work. Perhaps not surprisingly she found that different populations of the same species were being monitored in different locations using different ecological criteria. Information on the 172 monitoring projects came from 71 contacts and represented a good sample from the Natural Heritage Programmes and the Nature Conservancy Programmes. In addition to claims of no monitoring by 25 per cent of all contacts, three basic types of monitoring programmes for rare plants were identified; inventory, survey and demographic.

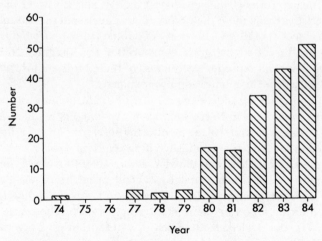

Fig. 16.4. The number of rare plant monitoring projects initiated in the USA from 1974 to 1984. From Palmer (1987).

The inventory method used was simply a population count repeated at intervals, usually each year. The survey method was based on transect or grid sampling so as to make estimates about the population characteristics or to obtain data on the entire population. The demographic studies involved marking or mapping individual plants in order to follow the status of those individuals.

The costs involved in these monitoring programmes may sometimes be seen as a trade-off in terms of the effectiveness of that monitoring method. Palmer (1986, 1987) noted for example that although an inventory is simple and not time consuming, data on reproductive biology may be overlooked and this could lead to erroneous conclusions about the status of a species. Palmer's critical examination of 172 rare plant monitoring projects in the USA led her to suggest 14 recommendations for rare plant monitoring (Table 16.2); recommendations which realistically consider biological parameters as well as logistic parameters.

One particularly interesting application of species monitoring is that undertaken to ensure that management practices on natures reserves do not harm those qualities which originally led to designation as a nature reserve. For example, of the monitoring programmes promoted by The Nature Conservancy in the USA, there are some which are directed at fulfilling legal obligations (S. C. Buttrick, pers. comm.). Occasionally the Conservancy leases rights on a nature reserve such as for grazing or for hay. Baseline data and then monitoring programmes are established to ensure that the activities do not harm the protected species. In some instances, such as on the Katharine Ordway Sycan Marsh Reserve in Oregon, Black Terns (*Chlidonias nigra*) require surface water of varying characteristics throughout their reproductive cycle and the Sandhill Crane (*Grus canadensis*) requires a diversity of habitats ranging from open water for nesting to tufted hairgrass communities for foraging. Number of breeding pairs and reproductive success of these birds are just two of the variables used in the monitoring programmes.

The importance of and relevance of rare plant monitoring became a much discussed issue in North America in the mid-1980s and some of the techniques used for rare plant monitoring in North America have been discussed in several papers published in the *Natural Areas Journal*, see for example Palmer (1986) and Bowles *et al.* (1986). A need for better communication was then noted by several members of the Ecological Society of America and as a result the *Natural Areas Journal* agreed to become a forum for a rare plant monitoring network. The centre for that network is the Holcomb Research Institute, Butler University (see Appendix II).

Table 16.2. *Palmer's (1987) recommendations for rare plant monitoring*

Planning stages
(1) Choose species or communities whose degree of rarity or endangerment warrant monitoring (Palmer 1986).
(2) Conduct preliminary sampling to determine an adequate number, size and shape of plots or transects needed to estimate parameters at the site (Green 1979).
(3) Describe results of preliminary sampling, goals and methods in a proposal for external review by experts in designs for field research. Propose the method of data analysis to ensure that the methods are appropriate for the plant.

Use of monitoring types
(4) Choose the inventory method for populations under no current threat in order to follow plant numbers over time and to make a preliminary assessment of need for further work.
(5) Choose the survey method for species under no current threat to follow such indicators of performance as the number of reproductive plants in the entire population.
(6) Choose the survey method to estimate numbers in very large populations or communities. Follow numbers in stage classes (seedlings, juveniles, adults) if possible.
(7) Choose the demographic method for sparse or threatened populations.
(8) Record plant sizes and follow all stage classes, including seedlings, in demographic studies. Use a stage-class transition matrix to identify critical stages and to predict changes in the population resulting from their management.
(9) Choose the demographic method when little is known about the life history of the species, or when detailed information about individual location, size or reproduction is otherwise pertinent to the project goals.
(10) Employ experimental manipulation in demographic or survey methods to test management treatments when there is a suspected threat to the site or to a life stage.

Sampling and experimental design
(11) Randomize sampling within each stratum in the habitat in survey and demographic studies.
(12) Replicate experimental units in studies that compare or test management procedures.
(13) Employ a design that acknowledges constraints in randomization when the sampling is not totally randomized.

Publications and reports
(14) Write proposals for review and annual progress reports. Ultimately publish the results in a journal.

World Conservation Monitoring Centre

The IUCN Red Data Books, after many years, continue to retain their immense value as a source of information but the system for gathering, checking and processing information from compilers around the world was not effective as it could be. The Books were also limited to a few groups; mammals, birds, amphibians and reptiles, fish, and flowering plants.

The Red Data Books were really a 'shop window' for much larger stores of information and there came a time when information technology was needed to improve the system. With the establishment of the World Conservation Monitoring Centre (previously the IUCN Conservation Monitoring Centre) in 1983 (Fig. 16.5), databases using sophisticated computerized systems soon improved the system for gathering, checking, storing and communicating information on plant and animal species throughout the world (Pellew 1991).

The animal database includes not only data on endangered species but also those depleted species of economic importance and also wild relatives of domestic stock. Data on single species has been the main format but in addition the database holds information on a regional basis such as the Banks Peninsula of New Zealand.

The WCMC plant database, with its information on 30 000 plant taxa has become a major source of information with which to operate a monitoring programme. The information available goes beyond that for specific plant taxa and can be used to answer such questions as 'are any wild forms of the potato endangered?' or 'which plant species are under threat in Poland?' (IUCN 1986).

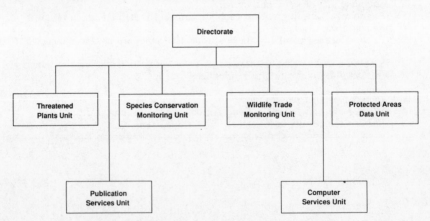

Fig. 16.5. The structure of the World Conservation Monitoring Centre, based at Cambridge and at Kew, England.

Monitoring the status of a species can be based on a subjective assessment of the information available in the database or alternatively precise ecological parameters can be used. However, the information about each species in the databases is not uniform and therefore the precise ecological parameters for monitoring a species will vary. For some species, monitoring may be based on numbers, for others it may be based on life tables and levels of recruitment.

Effective population size and genetic variation

One of the effects of insularization is isolation of populations (Chapter 13) and this in turn can lead to reduced genetic variability. Alternatively, populations of some species have been exploited to the extent that only small isolated populations remain; taken into captivity the genetic variation diminishes. There is a growing number of examples of species with populations so small that genetic variation has been severely depleted. The rhinoceroses of Africa and India (Merenlender *et al.* 1989) and Cheetahs in captivity (O'Brien *et al.* 1985) are just two large mammals in which genetic variation has been measured and shown to be very small. The implications are that these populations may not be able to adapt to changing environmental conditions.

The quantity of genetic variation in a population is a function of both the size of the population and the type of genetic variation but Lande & Barrowclough (1987) have described how it is possible to monitor the genetics of managed population. For example they described the use of structural gene polymorphisms as revealed by electrophorectic techniques. Assuming Hardy–Weinberg frequencies in the sample population, Lande & Barrowclough propose the following statistics which could be used to monitor variability.

Let x_{ij} be the frequency of allele i at locus j. Single-locus heterozygosity at locus j can then be estimated from the following (Lande & Barrowclough 1987):

$$h_j = 2n_j \left(1 - \sum_i x_{ij}^2 \right) \bigg/ (2n_j - 1)$$

where n_j is the number of individuals examined for locus j, and the sum is over all alleles segregating at that locus. Overall heterozygosity, H, is estimated as

$$H = (1/L) \sum_j h_j$$

where L is the total number of loci sampled, and the summation is over all loci $j = 1$ to L.

Alternatively, as done by Merenlender *et al.* (1989) in their study of genetic variation in rhinoceroses, it is possible to calculate heterozygosity directly from observed frequencies of actual heterozygotes at each locus.

Minimum populations size for conservation

Monitoring and managing populations is a challenge which will become more frequent as populations of more and more species become reduced, fragmented and isolated. Conservation of these reduced populations is made no easier by an all too often lack of knowledge on the genetics, ecology and behaviour of the speies.

It has long been known that as populations become reduced in size, they are more prone to extinction. Clearly there is a minimum population size, below which the chances of extinction become very great indeed. Shaffer (1981) provides us with a useful, tentative definition of minimum viable population size; 'the smallest isolated population having a 99 per cent chance of remaining extant for 100 years despite the foreseeable effects of demographic, environmental, and genetic stochasity, and natural catastrophes'. The per cent or the time span requires qualification depending on the life history strategy of the species but what is important is that such a concept illustrates an important objective of monitoring which was mentioned at the start of this book; that is, monitoring to determine the natural population dynamics of a species.

I introduced this book with a reference to census work on some Antarctic birds, census work which at that time seemed so easy with the use of a few simple ecological parameters. Perhaps the status of some species can be monitored successfully with simple methods and basic facilities but such an approach does little justice to the considerable importance of species monitoring. The monitoring of species for conservation has, over the last few decades, become a sophisticated science with an underlying sense of urgency. Furthermore, there are two extremes to species monitoring, one at the individual species level and one at the collective species level. That is in addition to the need for more species monitoring programmes there is a need for much greater support for monitoring the decline in numbers of species and that support could come from newly established biodiversity centres. Never before has the rate of species extinctions been so great, minimum estimates being at least one per day. Other estimates put levels of extinction in the order of several species a day (Prance 1991). Perhaps one benefit of species extinction monitoring programmes might be to draw attention to a phenomenon considered by some to be more important than world pollution.

References

Armstrong, D. W. & Cook, R. M. (1986). Proposal for a revised use of IYFS indices for calibrating VPA. ICES CM 1986/G, 3.

Bonner, W. N. (1986). The future of Antarctic resources. *The Geographical Journal*, 152, 248–55.

Borodin, A. M. (1978). *Red Data Book of USSR*. Moscow, Lesnaya Promyshlenost Publishers.

Bowles, M. L., Hess, W. J., DeMauro, M. M. & Hiebert, R. D. (1986). Endangered plant inventory and monitoring strategies at Indiana Dunes National Lakeshore. *Natural Areas Journal*, 6, 18–26.

Bullard, E. R., Shearer, H. D. H., Day, J. D. & Crawford, R. M. M. (1987). Survival and flowering of *Primula scotica* Hook. *Journal of Ecology*, 75, 589–602.

Council of Europe (1977). *List of Rare, Threatened and Endemic Plants in Europe*. Kew, Surrey, Nature and Environment Series, No. 14.

Curtis, T. G. F. & McGough, H. N. (1988). *The Irish Red Data Book*. Dublin, The Stationary Office.

Davy, A. J. & Jefferies, R. L. (1981). Approaches to the monitoring of rare plant population. In *The Biological Aspects of Rare Plant Conservation*, ed. H. Synge, pp. 219–32. Chichester, John Wiley.

Elliott, J. M. (1984). Numerical changes and population regulation in young migratory trout *Salmo trutta* in a Lake District stream, 1966–1983. *Journal of Animal Ecology*, 53, 979–94.

Elliott, J. M. (1988). Growth, size, biomass and production in contrasting populations of trout, *Salmo trutta*, in two Lake District streams. *Journal of Animal Ecology*, 57, 49–60.

Fairbrothers, D. E. & Hough, M. Y. (1975). Rare or endangered vascular plants of New Jersey. *New Jersey State Museum Scientific Notes*, 14, 1–53.

Gill, D. E., Berven, K. A. & Mock, B. A. (1983). The environmental component of evolutionary biology. In *Population Biology, Retrospect and Prospect*, ed. C. E. King & P. S. Dawson, New York, Columbia University Press.

Given, D. R. (1983). *Conservation of Plant Species and Habitats. A Symposium Held at 15th Pacific Science Congress, Dunedin, New Zealand, February 1983*. Nature Conservation Council, New Zealand.

Given D. R. (1988). *Monitoring threatened plants*. DSIR Resource Research, 14 December, 1988.

Given, D. R. (1989). Monitoring of threatened plants. In *Proceedings of a Symposium on Environmental Monitoring in New Zealand with Emphasis on Protected Natural Areas*, ed. by B. Craig, pp. 192–8, Wellington, Department of Conservation.

Green, R. H. (1979). *Sampling Design and Statistical Methods for Environmental Biology*. New York, John Wiley.

Hall, A. V., de Winter, B. & Oosterhout, S. A. M. (1980). *Threatened Plants of Southern Africa*, Pretoria, South African national Scientific Programmes Report No. 45. Cooperative Sciences Programme, Council for Scientific and Industrial Research.

Henderson, P. A. & Holmes, R. H. A. (1987). On the population biology of the common shrimp *Crangon crangon* (L.) (Crustacea: Caridea) in the Severn Estuary and Bristol Channel. *Journal of the Marine Biological Association*, 67, 825–47.

Henderson, P. A. & Holmes, R. H. A. (1989). Whiting migration in the Bristol Channel: a predator-prey relationship. *Journal of Fish Biology*, **34**, 409–416.

Henderson, P. A., Turnpenny, A. W. H. & Bamber, R. N. (1984). Long-term stability of a sand smelt (*Atherina presbyter* Cuvier) population subject to power station cropping. *Journal of Applied Ecology*, **21**, 1–10.

Hutchings, M. J. (1987). The population biology of the early spider orchid, *Ophrys sphegodes* Mill. 1. A demographic study from 1975 to 1984. *Journal of Ecology*, **75**, 711–27.

IUCN (1986). CMC – *its Work and Funding Requirements 1987–1989*. Gland, International Union for Conservation of Nature and Natural Resources.

Kershaw, L. J. & Morton, J. K. (1976). Rare and potentially endangered species in the Canadian flora – a preliminary list of vascular plants. *Canadian Botanical Association Bulletin*, **9**, 26–30.

Klimetzek, D. (1990). Population dynamics of pine feeding insects: a historical study. In '*M' Population Dynamics of Forest Insects*, ed. A. D. Watt, S. R. Leather, M. D. Hunter & N. A. C. Kidd. Andover, UK, Intercept.

Lande, R. & Barrowclough, G. F. (1987). Effective population size, genetic variation, and their use in population management. In *Viable Populations for Conservation*, ed. M. E. Soule, pp. 87–123. Cambridge, Cambridge University Press.

Leigh, J. & Boden, R. (1979). Australian flora in the endangered species convention CITES. *Australian National Parks and Wildlife Service Special Publication*, **3**, 93 p.

Malyshev, V. L. & Soboleuska, K. A. (1980). *Rare and Endangered Plant Species of Siberia*. (In Russian.) Hayka.

Merenlender, A. M., Woodruff, D. S., Ryder, O. A., Kock, R. & Vahala, J. (1989). Allozyme variation and differentiation in African and Indian Rhinoceroses. *Journal of Heredity*, **80**, 377–82.

Moss, R. & Watson, A. (1985). Adaptive value of spacing behaviour in population cycles of red grouse and other animals. In *Behavioural Ecology*, ed. R. M. Sibly & R. H. Smith, pp. 275–94. Oxford, Blackwell Scientific Publications.

O'Brien, S. J., Roelke, M. E., Marker, L., Newman, A., Winkler, C. A., Meltzer, D., Colly, L., Evermann, J. F., Bush, M. & Wildt, D. E. (1985). Genetic basis for species vulnerability in the cheetah. *Science*, **227**, 1428–34.

Palmer, M. E. (1986). A survey of rare plant monitoring: programs, regions, and species priority. *Natural Areas Journal*, **6**, 27–42.

Palmer, M. E. (1987). A critical look at rare plant monitoring in the United States. *Biological Conservation*, **39**, 113–27.

Peet, R. K. & Christensen, N. L. (1987). Competition and tree death. *BioScience*, **37**, 586–96.

Pellew, R. (1991). Data Management for Conservation. In *The Scientific Management of Temperate Communities for Conservation*, ed. I. F. Spellerberg, F. B. Goldsmith & M. G. Morris, pp. 505–22. Oxford, Blackwell Scientific Publications.

Perring, F. H. & Farrell, L. (1983). *British Red Data Books, 1, Vascular Plants*, 2nd ed. Nettleham, RSNC.

Peterson, R. O. (1987). *Ecological Studies of Wolves on Isle Royale, Annual Report – 1986–1987*. Houghton, MI, Michigan Technical University.

Prance, G. T. (1991). Rates of Loss of Biological Diversity: a Global Vie···. In *The Scientific Management of Temperate Communities for Conservation*, ed. I. F.

Spellerberg, F. B. Goldsmith & M. G. Morris, pp. 27–44, Oxford, Blackwell Scientific Publications.

Risser, P. G., Birney, E. C., Blocker, S. W., May, S. W., Parton, W. J. & Wiens, J. A. (1981). *The True Prairie Ecosystem.* Stroudsburg, PA, Hutchinson Ross Pub. Co.

Shaffer, M. L. (1981). Minimum population sizes for species conservation. *BioScience*, **31**, 131–4.

Tamm, C. O. (1972). Survival and flowering of some perennial herbs. 11. The behaviour of some orchids on permanent plots. *Oikos*, 23–8.

Taylor, L. F. (1986). Synoptic dynamics, migration and the Rothampsted Insect Survey. *Journal of Animal Ecology*, **55**, 1–38.

WCED (1987). *Our Common Future. World Commission on Environment and Development.* Oxford, New York, Oxford University Press.

Wells, T. C. E. (1981). Population ecology of terrestrial orchids. In *The Biological Aspects of Rare Plant Conservation*, ed. H. Synge, pp. 281–95. Chichester, Wiley.

Williams, G. R. & Given, D. R. (1981). *The Red Data Book of New Zealand.* Wellington, Nature Conservation Council.

Appendix I

Construction of a dendrogram

The following example refers back to Chapter 7 on the use of similarity indices, especially Savage's (1982) similarity index.

In Savage's account of the use of corixids in the classification of lakes, he used his indices of similarity to construct a dendrogram, based on a method used by Mountford (1962).

The following example shows how the four categories of sites on the right side of Fig. A1 were derived; viz. Esthwaite (E); Llyn Coron and Llyn Hendref (W); North West Midlands meres, group A (A); and North West Midlands meres, group B (B) (Savage 1982). The dendrogram for the last two groups has been published elsewhere (Savage & Pratt 1976). The indices of the four categories (sites) are set out in tabular form:

	E	W	A	B
E		0.524	0.600	0.322
W			0.928	0.847
A				0.750
B				

The highest index is selected from the table and the pair of sites (A and W) corresponding with this index are combined to form a single group (AW). The indices between this new group and each remaining site or group are then calculated:

$$I(E:AW) = \frac{I(E:A) + I(E:W)}{2} = \frac{0.600 + 0.524}{2} = 0.562$$

and

$$I(B:AW) = \frac{I(B:A) + I(B:W)}{2} = \frac{0.750 + 0.847}{2} = 0.798$$

A reduced table is made:

	E	AW	B
E		0.562	0.322
AW			0.798
B			

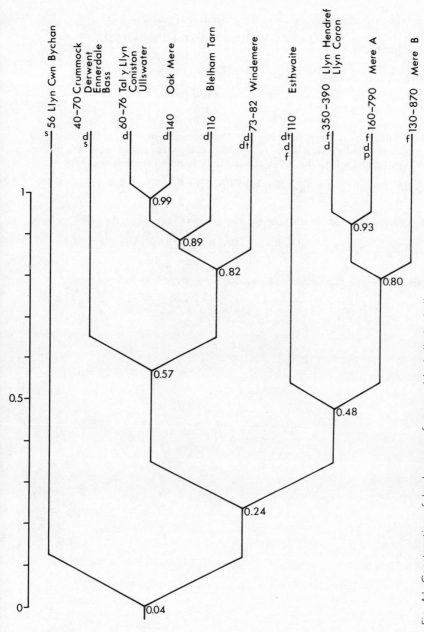

Fig. AI. Construction of dendrograms for use with similarity indices.

The highest index is selected and the procedure repeated:

$$I(E:AWB) = \frac{I(E:A) + I(E:W) + I(E:B)}{3} = \frac{0.600 + 0.524 + 0.322}{3} = 0.482$$

The final table is thus:

	E	AWB
E		0.482
AWB		

The dendrogram is drawn as in Fig. A1. The procedure may be further illustrated algebraically for five sites; A_1 A_2 A_3 B_1 and B_2 comprising two groups with indices $I(A_1A_2A_3)$ and $I(B_1B_2)$. The index of similarity between these two groups is

$$I(A_1A_2A_3:_1B_2) = \frac{I(A_1:B_1) + I(A_1:B_2) + I(A_2:B_1) + I(A_2:B_2) + I(A_3:B_1) + I(A_3:B_2)}{3 \times 2}$$

The general equation for the index of two groups of sites $A_1 \ldots A_m$ and $B_1 \ldots B_n$ is:

$$I(A_1 \ldots A_m:B_1 \ldots B_n) = \frac{I(A_1:B_1) + \ldots I(A_1:B_n) + \ldots I(A_m:B_1) + \ldots I(A_m:B_n)}{mn}$$

This is only one of a number of possible methods of calculation.

Appendix II

Organizations

The following organizations (together with acronyms) have been mentioned in the text either in relation to monitoring or in relation to materials and equipment for monitoring. This list is not claimed to be comprehensive. More comprehensive lists of environmental organizations may be found in Trzyna, T. C. (1989) *World Directory of Environmental Organizations*, 3rd ed., California Institute of Public Affairs and also in Barker, M. J. C. (1984) *Directory for the Environment*, Routledge & Kegan Paul, London. At the time of writing, a number of organizations are going through a period of reorganization or restructuring but the most recent address available has been provided.

AFRC
Agriculture and Food Research Council,
(Incorporating Rothampsted Experimental Station, Long Ashton Research Station and Broom's Barn Research Station)
Harpenden,
Herts AL5 2JQ,
UK.

Battelle Pacific Northwest Laboratories,
Richland,
Washington,
USA.

Biological Monitoring Inc.,
Blacksburg,
Virginia,
USA.

BRISC
Biological Recording in Scotland Campaign,
c/o the Scottish Wildlife Trust,
25 Johnston Terrace,
Edinburgh, EH1 1NH, Scotland.

BSBI
Botanical Society of the British Isles,
Department of Botany,
British Museum (Natural History),
Cromwell Rd.,
London SW7 5BD,
UK.

BAS
British Antarctic Survey,
High Cross,
Madingley Rd.,
Cambridge CB3 0ET,
UK.

British Butterfly Conservation Society,
Tudor House,
Quorn,
Loughborough,
Leicester LE12 8AD.
UK.

BES
British Ecological Society,
Burlington House,
Piccadilly,
London W1V 0LQ,
UK.

British Herpetological Society,
c/o Zoological Society of London,
Regent's Park,
London NW1 4RY,
UK.

British Lichen Society,
c/o Department of Botany,
British Museum (Natural History),
Cromwell Rd.,
London SW7 5BD,
UK.

BTO
British Trust for Ornithology
Beech Grove,
Station Rd.,
Tring,
Herts HP23 5NR,
UK.

BP International
Biological Services Division,
Britannic House,

Moor Lane,
London EC2Y 9BU,
UK.

Centre Recherches sur la Biologie des Populations d'Oiseaux (CRBPO),
55, rue de Buffon,
75005 Paris,
France.

CONCAWE
Conservation of Clean Air and Water in Western Europe
Babylon Kantoren A. Koningen Julianaplein 30–9,
2595 AA,
The Hague,
The Netherlands.

Convention on International Trade in Endangered Species,
(CITES) Secretariat,
6 Rue Maupas,
Case Postale 78,
CH-1,000,
Lausanne 9,
Switzerland.

Countryside Commission
John Dower House,
Crescent Place,
Cheltenham,
Glos. GL50 3RA,
UK.

DOE
Department of the Environment,
2 Marsham St.,
London SW1P 3EB,
UK.

Ecological Society of America,
Department of Biology,
Notre Dame University,
Notre Dame,
IN 46556,
USA.

EPA
Environmental Protection Agency,
401 Street,
Washington,
DC 20460,
USA.

Fair Isle Bird Observatory Trust,
21 Regent Terrace,
Edinburgh EH7 5BT,
Scotland.

FAO
Food and Agricultural Organization,
Viale delle Terme di Caracalla,
Rome,
Italy.

Farmlink,
(Farm Management Information),
Mendip House,
High St,
Taunton,
Somerset TA1 3SX,
UK.

FSC
Field Studies Council,
Preston Montford,
Montford Bridge,
Shrewsbury SY4 1HW,
UK.

FSCRC
Field Studies Council Research Centre,
(Previously OPRU – Oil Pollution Research Unit),
Fort Popton,
Angle,
Dyfed SA71 5AD,
UK.

FBA
Freshwater Biological Association,
The Ferry House,
Far Sawrey,
Ambleside,
Cumbria LA22 0LP,
UK.

FoE
Friends of the Earth,
International Secretariat,
Nicole Mueller,
26–28 Underwood St,
London N1 7JQ,
UK.

Gallamann Memorial Foundation,
PO Box 45593,
Nairobi,
Kenya.

The Game Conservancy,
Burgate Manor,
Fordingbridge,
Hampshire SP6 1EF,
UK.

GEMS & EARTHWATCH
Global Environmental Monitoring System,
PO Box 30552, Nairobi,
Kenya.

GeoData Institute,
Southampton University,
Southampton SO9 5NH,
UK.

GIS Oiseaux Marins,
55 rue Buffon,
75005 Paris,
France.

Holcomb Research Institute,
Butler University,
Indianapolis,
IN 46208,
USA.

Institute of Freshwater Ecology,
Windermere Laboratory,
The Ferry House,
Ambleside,
Cumbria LA22 0LP,
UK.

ITE
The Institute of Terrestrial Ecology,
Bush Estate,
Penicuik,
Midlothian EH26 0QB,
UK.

The ITE Predatory Birds Monitoring Scheme,
Monks Wood Experimental Station,
Abbots Ripton,
Huntingdon PE17 2LS,
UK.

IEEP
Institute for European Environmental Policy,
(1) Aloys-Schulte-Strasse 6,
BRD-5300 Bonn 1,
Germany.

(2) 3 Endsleigh St,
London WC1H 0DD,
UK.

ICBP
International Council for Bird Preservation,
219c Huntingdon Rd,
Cambridge CB3 0DL,
UK.

ICSU
International Council of Scientific Unions,
51 boulevard de Montmorency,
F-75016,
France.

INFOTERRA
International Referral System for Sources of Environmental Information.
See UNEP.

IUBS
International Union of Biological Sciences,
Dr Tala Younes,
51 Bd. de Montmorency,
F-75016,
Paris,
France.

IUCN
International Union for the Conservation of Nature and Natural Resources
(1) IUCN Communications Unit,
Avenue du Mont-Blanc,
CH-1196 Gland,
Switzerland.

(2) IUCN Publications Services Unit,
219c Huntingdon Rd,
Cambridge CB3 0DL,
UK.

Irish Wildbird Conservancy,
Ruttledge House,
8 Longford Place,
Monkstown,
Co. Dublin,
Ireland.

IWC
International Whaling Commission,
The Red House,
Station Rd,
Histon,
Cambridge CB4 4NP.
UK.

Land Use Research Unit,
Geography Department,
King's College,
Strand,
London WC2R 2LS,
UK.

Mammal Society,
c/o The Linnean Society,

Burlington House,
Piccadilly,
London w1v 0lq,
UK.

MARC
Monitoring and Research Centre,
King's College London,
University of London,
UK.

Mitre Corporation,
Burlington Rd,
Bedford,
MA 01730,
USA.

NASA
National Aeronautics and Space Administration,
600 Independence Avenue,
SW, Washington,
DC 20546,
USA.

National Programme for Ecosystem Research,
Council for Scientific and Industrial Research,
Pretoria,
South Africa.

National Rivers Authority,
Rivers House,
30–40 Albert Embankment,
London se1 7tl,
UK.

Norwich Union Costwatch UK,
Department of Science and Environmental Technology,
Farnborough College of Technology,
Farnborough,
Hampshire gu14 6sb,
UK.

NSF
National Science Foundation,
Washington,
DC 20550,
USA.

The Nature Conservancy,
1815 North Lynn St,
Arlington,
Virginia 22209,
USA.

NCC
Nature Conservancy Council,
Northminster House,
Peterborough PE1 1UA,
UK.

National Audubon Society,
950 Third Ave,
New York,
NY 10022,
USA.

National Swedish Environment Monitoring Programme (PMK),
National Environment Protection Board,
P.O. Box 1302,
S-171 25 Solna,
Sweden.

The National Wildlife Federation,
1400 16th St,
NW, Washington,
DC,
USA.

NERC
Natural Environment Research Council,
Polaris House,
North Star Avenue,
Swindon,
Wiltshire SN2 1EU,
UK.

NERC Environmental Information Centre,
The Institute of Terrestrial Ecology,
Monks Wood Experimental Station,
Abbots Ripton,
Huntingdon PE17 2LS,
UK.

New York Botanical Garden,
Mary Flagler Cary Arboretum,
Box AB,
Millbrook,
NY 12545,
USA.

NUTIS
Natural Environment Research Council,
Unit for Thematic Information Systems,
Department of Geography,
University of Reading,
P.O. Box 227,
Reading RG6 2AB,
UK.

Peoples Trust for Endangered Species,
Hamble House,
Meadrow,
Godalming,
Surrey GU7 3JX,
UK.

Philip Harris Biological Ltd.,
Oldmixon,
Weston-super-Mare,
Avon BS24 9BJ,
UK.

Plymouth Marine Laboratory (NERC),
Prospect Place,
The Hoe,
Plymouth PL1 3DH,
UK.

Research Institute for Nature Management (Netherlands),
Postbus 46,
3956 2R Heersum,
The Netherlands.

RSPB
Royal Society for the Protection of Birds,
The Lodge,
Sandy,
Beds. SG19 2DL,
UK.

SCAR
Scientific Committee on Antarctic Research
(one of ICSU units).

SCOPE
National Committee on Problems of the Environment
(one of the ICSU units).

Sea Mammal Research Unit,
c/o British Antarctic Survey,
High Cross,
Madingley Rd,
Cambridge CB3 0ET,
UK.

Sierra Club,
530 Bush St,
San Francisco,
CA 94108,
USA.

Smithsonian Institution,
1000 Jefferson Drive,
SW, Washington,
DC 20520,
USA.

SOVON (Netherlands),
Postbus 81,
6573 2H Beek,
The Netherlands.

Swedish Ornithological Society,
(Sveriges Ornitologiska Forening).
Box 14219,
10440 Stockholm,
Sweden.

UNEP
United Nations Environment Programme,
P.O. Box 30552,
Nairobi,
Kenya.

UNESCO
UNESCO House,
7 Place de Fontenoy, 75700,
Paris,
France.

US Environment Protection Agency,
401 M Street,
SW Washington,
DC 20460,
USA.

US Fish and Wildlife Service,
Dept of the Interior,
Washington,
DC 20240,
USA.

WATCH Trust for Environmental Education Ltd.,
22 The Green,
Nettleham,
Lincoln LN22 2NR,
UK.

WCED
World Commission on Environment and Development,
Palais Wilson,
52 rue des Pagnis,
Ch-1001,
Geneva,
Switzerland.

WHO
World Health Organization,
1211 Geneva 27,
Switzerland.

Wild Bird Society of Japan,
Aoyama Flower Bldg,
1-1-4 Shibuya,
Shibuya-ku,
Tokyo 150,
Japan.

Wildfowl and Wetlands Trust,
Gatehouse,
Slimbridge,
Glos. GL2 7BT,
UK.

Wildlife Link,
45 Shelton St,
London, WC2H 9HJ,
UK.

World Conservation Monitoring Centre,
219c Huntingdon Rd,
Cambridge CB3 0DL,
UK.

World Resources Institute
1709 New York Avenue,
NY,
Washington,
DC,
USA.

WWF (International)
Avenue du Mont-Blanc,
CH-1196, Gland,
Switzerland.

WWF (UK)
World Wide Fund for Nature,
Panda House,
Ockford Rd,
Godalming,
Surrey GU7 1QU,
UK.

WWF (USA)
1250 24th St.
NW, Washington
DC 200337
USA.

Appendix III

Selected sources of information for identification of taxa

Biological and ecological monitoring is impossible without identification of organisms and therefore it seems useful to give an introduction to suitable literature. This is indeed an introduction (and no more than that) to some literature about identification of living organisms or books which lead to taxonomic literature. It is not practical to provide a comprehensive list because there is a vast amount of literature available; for example, plants of many regions are described in either large, definitive publications such as *Flora Europaea* or in smaller pictorial books such as Vedel (1978) *Trees and Shrubs of the Mediterranean*. This list is therefore meant to give an indication only of the kind of literature available for some regions of the world and it is obvious that the list is limited to a small number of taxonomic groups. The references have been grouped under zoogeographical regions.

International

Blake, S. F. & Atwood, A. C. (1967). *Geographical Guides to the Floras of the World.* US Dept. of Ag. Washington, D.C.

Davis, E. B. (1987). *Guides to Information Sources in the Botanical Sciences.* Littleton, Col. Libraries Unlimited.

Drebes, G. (1974). *Marine Phytoplankton,* Stuggart, George Thieme Verlag.

Frodin, D. G. (1985). *Guide to the Standard Floras of the World.* Cambridge, Cambridge University Press.

Gilbert, P. & Hamilton, C. J. (1983). *Entomology. A Guide to Information Sources.* London, Mansell Pub. Ltd.

Nelson, J. S. (1984). *Fishes of the World.* New York, Chichester, John Wiley.

Tuck, G. S. & Heinsel, H. (1978). *A Field Guide to the Seabirds of the World.* London, Collins.

Australasia

There is a *Flora of Australia,* and the *Flora of New Zealand* is being published in several volumes over a period of years.

A. Series, monographs and synopses

DSIR (New Zealand). The DSIR has published keys and guides for many of New Zealand's fauna.

Mobil Oil Co. A series of field guides for the fauna and flora of New Zealand has been sponsored by the Mobil Oil Company.

B. General references

Ayling, T. & Cox, G. J. (1982). *Collins Guide to the Sea Fishes of New Zealand.* Collins.

Bradstock, M. (1985). *Between the Tides: New Zealand Shore and Estuarine Life.* Reed Methuen.

Common, I. F. B. & Waterhouse, D. F. (1972). *Butterflies of Australia.* Angus & Robertson.

Cronin, L. (1988). *Key Guide to Australian Trees.* Reed.

Falla, R. A., Sibson, R. B. & Turbot, E. G. (1989). *A Field Guide to the Birds of New Zealand.* Collins.

Gill, B. (1986). *Collins Handguide to the Frogs and Reptiles of New Zealand.* Auckland, Collins.

Goode, J. (1980). *Insects of Australia.* Angus & Robertson.

Grigg, J. (1977). *Insects. Science Field Guides.* Sydney, Reed Education.

Gunson, D. (1983). *Collins Guide to the New Zealand Seashore.* Auckland, Collins.

Harvey, M. E. & Yen, A. (1989). *Worms to Wasps. Australia's Terrestrial Invertebrates.* Oxford, Oxford University Press.

Hawksworth, T. (1987). *Beetles of Australia.* Angus Robertson.

McLay, C. (1988). *Crabs (Brachyura and crab-like Anomura) of New Zealand.* Leigh Laboratory Bulletin, 22.

Miller, D. (1952). *Common Insects in New Zealand.* Wellington, A. H. & A. W. Reed.

Moore, L. B. & Irwin, J. B. (1978). *The Oxford Book of New Zealand Plants.* Oxford, Oxford University Press.

Paulin, C., Stewart, A., Roberts, C. & McMillan, P. (1989). *New Zealand Fish. A Complete Guide.* National Museum of New Zealand Miscellaneous Series No. 19, LGP Books, Government Printing Office.

Pizzey, G. (1980). *A Field Guide to the Birds of Australia.* Collins.

Pollard, J. (1980). *Handbook of Australian Fishes.* J. Pollard, Ltd.

Powell, A. W. B. (1983). *New Zealand Mollusca.* Collins.

Smith, B. J. & Kershaw, R. C. (1979). *Field Guide to the Non-Marine Molluscs of South East Australia.* Canberra, ANU Press.

Oriental

There are many comprehensive Floras of this region such as: *FLORA Reipublicae Popularis Sinicae* (1988), Science Press for China; *Flora Malesiana*, published by the Rijksherberium, Leiden, Netherlands; *Flora of Java*, Groningen, Walters-Noordhoff.

General references

Ali, S. & Ripley, S. D. (1968). *Handbook of the Birds of India and Pakistan*, Ten Vols. Bombay, London, Oxford University Press.

Barlow, H. S. (1982). *An Introduction to the Moths of Southeast Asia*. Kuala Lumpur, Malayan Nature Society.

Corbetm, A. S. & Pendlebury, H. M. (1978). *The Butterflies of the Malay Peninsula*. Kuala Lumpur, Malaya Nature Society.

Hails, C. (1986). *Birds of Singapore*, Singapore, Times Editions.

Holloway, J. D. (1976). *Moths of Borneo with Special Reference to Mount Kinabalu*. Kuala Lumpur, The Malayan Nature Society.

Holttum, R. E. (1977). *Plant Life in Malaya*, 2nd ed. Kuala Lumpur, Longman.

King, B., Woodcock, M. & Dickinson, E. C. (1975). *A Field Guide to the Birds of South-east Asia*. London, Collins.

Lekagul, B. & McNeeley, J. A. (1977). *The Mammals of Thailand*. Association for the Conservation of Wildlife, Bangkok.

MacKinnon, J. (1988). *Field Guide to the Birds of Java and Bali*. Yogyakarta, Gadjah Mada University Press.

Payne, J. (1987). *A Field Guide to the Mammals of Borneo*. WWF, Kuala Lumpur and Sabah Society, Kola Kinabalu.

Roberts, D. (1982). *Shallow Water Marine Molluscs of North-West Java*. Jakarta, LIPI.

Tobing, D. H. (1968). *A Preliminary Bibliography on Fauna, Flora Vegetation Conservation of Nature and Natural Resources of Sumatra*. Nature Conservation Dept., Agricultural University, Wangeningen.

Whitmore, T. C. (1972). *Tree Flora of Malaya*. Kuala Lumpur, Longman.

Whitmore, T. C. (1977). *Palms of Malaya*. Kuala Lumpur, Oxford University Press.

Whitten, A. J., Damanik, S. J., Anwar, J. & Hisyam, N. (1984). *The Ecology of Sumatra*. Yogyakarta, Gadjah Mada University Press.

Whitten, A. J., Kottelat, M., Kartikasari, S. N., Wirjoatmodjo, S. & Yuwono, D. (1991). *The Fishes of Lakes and Rivers in Western Indonesia and Sulawesi*. Jakarta, Gramedia.

Whitten, A. J., Mustafa, M. & Henderson, G. (1987). *The Ecology of Sulawesi*. Yogyakarta, Gadjah Mada University Press.

Palaearctic

As well as a *Flora Europaea* and a *Flora USSR*, there are many other Floras of various regions in the Palaearctic.

A. Series, monographs and synopses

1. The Linnean Society of London, Burlington House, Piccadilly, London W1V 0LQ, UK.

The Linnean Society has published synopses of the British Fauna consisting of field and laboratory guides which are suitable both for the amateur and professional. There are currently 43 in the series which cover a wide range of invertebrate groups, both marine and terrestrial.

2. The Royal Entomological Society of London, 41 Queen's Gate, London SW7 5HU, UK.

The Royal Entomological Society of London has published a series of handbooks for the identification of British insects.

3. The Field Studies Council, Preston Montford, Montford Bridge, Shrewsbury, SY4 1HW, UK.

The Field Studies Council publishes AIDGAP-tested keys for the identification of groups of animals and plants for which there are no suitable keys. The journal *Field Studies*, published by the Field Studies Council, has included a number of keys.

4. Naturalists' Handbooks. Guides for the exploration of British natural history. Published by Richmond Publishing Co. Ltd.

5. Systematics Association (MAFF, Slough Laboratory, London Rd, Slough, Berks. SL3 7HJ, UK): see *Key Works* by Kerrich *et al.* (1978).

6. The Freshwater Biological Association (FBA).
The FBA has published a number of keys under the title of their Scientific Publications.

B. General references

Bishop, O. N. (1973). *Natural Communities*. London, John Murray.

Blanc, M., Banarescu, P., Gaudet, J.-L. & Hureau, J.-C. (1971). *European Inland Water Fish. A Multilingual Catalogue*. London, FAO and Fishing News Books.

Bunce, R. G. H. (1982). *A Field Key for Classifying British Woodland Vegetation*, Part 1. NERC.

Bursche, E. M. (1971). *A Handbook of Water Plants*, English Translation. London, Frederick Warne.

Clapham, A. R., Tutin, T. G. & Warburg, E. F. (1968). *Excursion Flora of the British Isles*. Cambridge, Cambridge University Press.

Hureau, J. C. & Monod, T. (1973). *Check-list of the Fishes of the North-Eastern Atlantic and of the Mediterranean* (CLOFNAM). UNESCO, Paris.

Jones-Walters, L. M. (1989). Keys to the families of British Spiders. *Field Studies*, **9**, 365–443.

Kerrich, G. J., Hawksworth, D. L. & Sims, R. W. (1978). *Key Works to the Fauna and Flora of the British Isles and North-Western Europe*. London, Academic Press.

Lockett, G. H. & Millidge, A. F. (1951). *British Spiders*, Vol. 1. London, The Ray Society.

Macan, T. T. (1959). *A Guide to Freshwater Invertebrate Animals*. Longman, London.

Muus, B. J. & Dahlstron, P. (1967). *Freshwater Fish of Britain and Europe*. London, Collins.

Sims, R. W. (1980). *Animal Identification. A Reference Guide*. Chichester, John Wiley.

Nearctic

A. Series, monographs and synopses

Wm. C. Brown Publishers (P.O. Box 539, Dubuque, IA 52001) have produced a series of illustrated keys and guides for a wide range of taxa, *PKN Series ... You know what it is.*

B. General references

Bacon, P. R. (1978). *Flora & Fauna of the Caribbean*. Trinidad, Key Caribbean Publications.

Brusca, R. C. (1980). *Common Intertidal Invertebrates of the Gulf of California.* University of Arizona Press.

Burt, W. H. & Grossenheider, R. P. (1976). *A Field Guide to the Mammals.* Boston, Houghton Mifflin Co.

Conant, I. H. (1977). *A Field Guide to Reptiles and Amphibians of Eastern and Central North America.* Boston, Houghton Mifflin Co.

Covell, C. V. (1984). *A Field Guide to the Moths of Eastern North America.* Boston, Houghton Mifflin.

Eddy, S. & Hodson, A. C. (1961). *Taxonomic Keys to the Common Animals of the North Central States Exclusive of the Parasitic Worms, Insects and Birds.* Minn. Burgess Publishing Co.

Ehrlich, P. R., Dobkin, D. S. & Wheye, D. (1988). *The Birder's Handbook.* New York and London, Simon & Schuster Inc.

Hurlbert, S. H. & Villalobos-Figueroa, A. (eds) (1982). *Aquatic Biota of Mexico, Central America and the West Indies.* San Diego, San Diego State University.

Kozloff, E. N. (1987). *Marine Invertebrates of the Pacific Northwest.* Washington, D.C., University of Washington Press.

Littler, D. S., Littler, M. M., Bucher, K. E. & Norris, J. N. (1989). *Marine Plants of the Caribbean.* Smithsonian Institution Press.

Morris, R. H., Abbott, D. P. & Haderlie, E. C. (1980). *Intertidal Invertebrates of California.* Stanford University Press.

National Geographic Society (1987). *Field Guide to the Birds of North America.* Washington, National Geographical Society.

Pennak, R. W. (1989). *Fresh-water Invertebrates of the United States: Protozoa to Mollusca.* New York, Wiley.

Petrides, G. A. (1958). *A Field Guide to Trees and Shrubs.* Boston, Houghton Mifflin Co.

Ricketts, E. F. & Calvin, J. *Between Pacific Tides,* 5th ed. Stanford University Press. (In press.)

Ruppert, E. E. & Fox, R. S. (1988). *Seashore Animals of the Southeast.* University of South Carolina Press.

Sefton, N. & Webster, S. (1986). *Caribbean Reef Invertebrates.* Sea Challengers, California, Monterey.

Spellenberg, R. (1979). *The Audubon Society Field Guide to North American Wildflowers.* New York, Alfred A. Knopf.

Swan, L. A. & Papp, C. S. (1972). *Common Insects of North America.* New York, Harper & Row.

Neotropical

General references

Barrington, H. C. (1974). *Medicinal de Colombia. Botanica Medica.* Bogota, Instituto de Cienias Naturales Universidad Nacionae.

Eisenberg, J. F. (1989). *Mammals of the Neotropics.* Chicago & London, University of Chicago Press.

Fittkau, E. J., Illies, J., Klinge, H., Schwabe, G. H. & Sioli, H. (eds) (1968). *Biogeography and Ecology in South America.* Two Vols. The Hague, Dr. W. Junk, N.V.

Harling, G., Sparre, B. & Anderson, C. (a series issued at indefinite intervals). *Flora of Ecuador.* Stockholm, NFR.

Meyer de Schauensee, R. (1970). *A Guide to the Birds of South America*. Edinburgh, Oliver & Boyd.

Meyer de Schauensee, R. (1964). *The Birds of Columbia and Adjacent Areas of South and Central America*. Narbeth, Pa., Livingstone Pub. Co.

Prance, G. T. & Lovejoy, T. E. (1985). *Key Environments. Amazonia*. Oxford, Pergamon Press.

Ethiopian

There is an *African Flora* as well as many regional guides to the flora.

General references

Branch, B. (1988). *Field Guide to the Snakes and other Reptiles of Southern Africa*. Cape Town, Struik Publishers.

Dirsch, V. M. (1965). *The African Genera of Acridoidea*. London, Cambridge Univesity Press.

IUCN/UNEP (1987). *The IUCN Directory of Afrotropical Protected Areas*. Gland, IUCN.

Keay, R. W. J. (1989). *Trees of Nigeria*. Oxford, Clarendon.

Kingdon, J. (1971). *Mammals of East Africa*. London, Academic Press.

Kingdon, J. (1990). *Island Africa*. London, Collins.

Lewis, A. & Pomeroy, D. (1989). *A Bird Atlas of Kenya*. Rotterdam, A.A. Balkema.

Lind, E. M. & Morrison, M. E. S. (1974). *East African Vegetation*. London, Longman.

Mackworth-Praed, C. W. & Grant, C. H. B. (1952). *African Handbook of Birds*, London, Longman.

Maclean, G. L. (1985). *Roberts' Birds of Southern Africa*. Cape Town, CTP Book Printers.

Migdoll, I. (1987). *Field Guide to the Butterflies of Southern Africa*, Struik, Cape Town.

Palgraves, K. C. (1977). *Trees of Southern Africa*. Cape Town, Struik, Publishers.

Passmore, N. I. & Carruthers, V. C. (1979). *South African Frogs*. Witwatersrand University Press, Johannesburg.

Scholtz, C. H. & Holm, E. (1985). *Insects of Southern Africa*. Butterworths, Durban.

Seret, B. (1981). *Poissons de Mer de L'Ouest Africain Tropical*. Paris, Orstom.

Skaife, S. H. (1979). *African Insect Life*. Cape Town, C. Struik Publishers (PTY) Ltd.

Smithers, R. H. (1983). *Mammals of the Southern African Subregion*. Pretoria, Pretoria University Press.

White, F. (1983). *The Vegetation of Africa: A Descriptive Memoir to Accompany the UNESCO/AERFAT/UNSO Vegetation Map of Africa*. Paris, UNESCO.

Williams, J. E. (1968). *A Field Guide to the Butterflies of Africa*. Collins, London.

Index

328